Forensic Entomology

An Introduction

Dorothy E. Gennard
University of Lincoln, UK

BICENTENNIAL
1807
WILEY
2007
BICENTENNIAL

Other Wiley Editorial Offices

John Wiley & Sons Inc., 111 River Street, Hoboken, NJ 07030, USA

Jossey-Bass, 989 Market Street, San Francisco, CA 94103-1741, USA

Wiley-VCH Verlag GmbH, Boschstr. 12, D-69469 Weinheim, Germany

John Wiley & Sons Australia Ltd, 33 Park Road, Milton, Queensland 4064, Australia

John Wiley & Sons (Asia) Pte Ltd, 2 Clementi Loop #02-01, Jin Xing Distripark, Singapore 129809

John Wiley & Sons Canada Ltd, 22 Worcester Road, Etobicoke, Ontario, Canada M9W 1L1

Wiley also publishes its books in a variety of electronic formats. Some content that appears in print may
not be available in electronic books.

Anniversary Logo Design: Richard J. Pacifico

Library of Congress Cataloging in Publication Data

Gennard, Dorothy E.
  Forensic entomology : an introduction / Dorothy E. Gennard.
    p. cm.
  Includes bibliographical references.
  ISBN: 978-0-470-01478-3 (cloth : alk. paper)
  ISBN: 978-0-470-01479-0 (pbk. : alk. paper)
  1. Forensic entomology.   I. Title.
  RA1063.45.G46 2006
  614′.1—dc22   2006032094

British Library Cataloguing in Publication Data

A catalogue record for this book is available from the British Library

ISBN 978-0-470-01478-3 (HB)   978-0-470-01479-0 (PB)

Typeset in 10.5/12.5pt Times by Integra Software Services Pvt. Ltd, Pondicherry, India
Printed and bound in Great Britain by Antony Rowe Ltd, Chippenham, Wiltshire
This book is printed on acid-free paper responsibly manufactured from sustainable forestry in which
at least two trees are planted for each one used for paper production.

Contents

List of figures

List of tables

Preface

This book is an introduction to forensic entomology for undergraduates, particularly those studying for a degree in forensic science. Responding to comments over the past 6 years from students studying forensic entomology, this book provides a basic entomological background, with descriptions of practical techniques and the legal aspects of using insects to estimate the time since death and help solve crime. I have included sections on the ecological implications of the presence of some of the more frequent insect visitors to a corpse (human and non-human). I have also included information from a range of countries to broaden the application of the textbook, as students travel widely and may go on to find employment across the world.

I hope that, by using *Forensic Entomology: An Introduction*, students will find studying entomology interesting and its role and application as a forensic tool to solve crime scenarios a fascination, irrespective of any fears about the smell!

Dorothy Gennard

Acknowledgements

I am grateful to the following for permission to reproduce illustrations, or quote from publications:

The Amateur Entomologists' Society, for permission to quote from Dear J. P. *Carrion* (1978). In Stubbs A. and Chandler P. *A Dipterist's Handbook*. The Amateur Entomologist 15. The Amateur Entomologists' Society: London (1978).

Dr Mark Benecke, for providing a copy of his RAPD profile of *Oiceophoma thoracicum* L. and *Calliphora vicina* (Robineau-Desvoidy).

Professor Michael Claridge, the Royal Entomological Society of London and Dr B. R. Laurence, for permission to reproduce his photograph of the tequila bottle from *Antenna* **6**(3) (1982).

Dr Jonathan Cooter, Hereford Museum, for permission to quote his comments about the distribution of *Necrobia ruficollis* in the UK.

The Etnografisch Museum of Antwerp, for providing a CD of photographs of the paintings by Morishige (1673–1680) entitled 'The nine contemplations of the impurity of the human body' (Figures 1.1–1.7) and for permission to reproduce them. ..

Dr Sharon Erzinçlioğlu, for permission to quote the case of Mike Evans and Zoë, from Erzinçlioğlu Y. Z. *Maggots, Murder and Men*. Harley Books: Colchester (2000).

Elsevier, for permission to quote details of lower temperature limits, published in Marchenko M. L. Medico-legal relevance of cadaver entomofauna for the determination of the time of death. *Forensic Science International* **120**(1–2): 89–109 (2001) and a section of Forensic Science International 98.

Benecke M., Random amplified polymorphic DNA (RAPD) typing of necrophageosis insects (Diptera: Coleoptera) in criminal forensic science: validation and use in practice (Figure 4) p. 164 © (1998) with permission from Elsevier (presented in this book as Figure 3.17).

Dr Susan Giles, Curator, Bristol Museums, Galleries and Archives, Bristol City Council, for the opportunity to photograph the Bristol Mummy.

Dr Martin Hall, British Museum, Natural History, for permission to quote his comments on the species of fly recovered from the bodies of two children murdered in Soham, Cambridgeshire.

Frances Harcourt-Brown, for permission to reproduce her photograph (Figure 1.11) of a rabbit exhibiting myiasis.

Dr Kate Horne, Secretary of the Council for the Registration of Forensic Practitioners, for permission to reproduce the tenets of the Council for the Forensic Practitioner.

The Regents of the University of California, for permission to reproduce a modification of Figure 4A from Wilson L. T. and Barnett W. W. Degree-days: an aid in crop and pest management. *California Agriculture* (**January–February**): 4–7 (1983). ™©1983 Regents, University of California (presented in this book as Figure 7.3).

John Newton and Rentokil Initial plc, for permission to reproduce diagrams of antennae, insect larvae and pupae (Figures 2.3, 3.4 and 3.5), from Munro J. W. *Pests of Stored Products*. The Rentokil Library. Benham and Co: Colchester (1965).

Tecknica Ltd, for providing the 2006 MapMate data and for permission to reproduce the Checklist of UK Recorded Calliphoridae, 2006, and also selected families of the Checklist of UK Coleoptera.

Dr Chris Pamplin, Editor, UK Register of Expert Witnesses, for permission to reproduce a section of *Taking Experts out of Court*, citing evidence for the Daubert test of evidence.

Warrant Officer Thierry Pasquerault Colonel J. Hebfard, of the Criminal Research Institute of the French and National Gendarmerie, for permission to use his photograph of packaging of entomological specimens recovered from a crime scene.

I would like to thank the following people for their comments, discussion and advice:

Mr Bill Barnett; Dr Trevor Crosby, Curator NZ Arthropod Collection, Landcare Research, University of Auckland; Keith Butterfield, University of Lincoln; Mr Lindsay Cutts; Dr John Esser; Susan Giles, Curator, Bristol Museums, Galleries and Archives, Bristol City Council; Katrina Hanley; Lorna Hanley; Karen Inckle; Mr Simon Jelf, Bond Solon Training Ltd; Helen Joiner; Mark Lawton; Helen Lonsdale; Dr Robert Nash, Ulster Museum; Professor Brian McGaw, University of Lincoln; Dr Brett Ratcliffe, Curator and Professor, Systematics Research Collections, University of Nebraska, Lincoln, NE, USA; Sean Riches, CAB Product Manager, Novartis Ltd, Paula Saward, formerly of Centrex; Kate Stafford; Dr J. Van Alphen, Etnografisch Museum of Antwerp, Belgium; Arpad Vass, Oakridge National Laboratory, TN, USA; Janet L. White,

Executive Editor, *California Agriculture*; Dr Lloyd T. Wilson, Texas A&M University; Laura Woodcock and Dr Frank Zalom, University of California.

I am most grateful to Rachael Ballard and Elizabeth Kingston, of John Wiley & Sons Ltd, for their help, encouragement and support; especially to Elizabeth Kingston and Lesley Winchester (Freelance Editor) for their editing assistance in the production of this manuscript.

I am grateful to Dr Darren Mann and Mr James Hogan of the Hope Entomological Collections, Oxford University Museum of Natural History, for providing, on extended loan, dipteran specimens for photography.

I am especially grateful to Mr David Padley, formerly police photographer with the Lincolnshire Police, for his excellent photography of the entomological specimens.

1

The breadth of forensic entomology

Forensic entomology is the branch of forensic science in which information about insects is used to draw conclusions when investigating legal cases relating to both humans and wildlife, although on occasion the term may be expanded to include other arthropods. Insects can be used in the investigation of a crime scene both on land and in water (Anderson, 1995; Erzinçlioğlu, 2000; Keiper and Casamatta, 2001; Hobischak and Anderson, 2002; Oliveira-Costa and de Mello-Patiu, 2004). The majority of cases where entomological evidence has been used are concerned with illegal activities which take place on land and are discovered within a short time of being committed. Gaudry *et al.* (2004) commented that in France 70 % of cadavers were found outdoors and of these 60 % were less than 1 month old.

The insects that can assist in forensic entomological investigations include blowflies, flesh flies, cheese skippers, hide and skin beetles, rove beetles and clown beetles. In some of these families only the juvenile stages are carrion feeders and consume a dead body. In others both the juvenile stages and the adults will eat the body (are necrophages). Yet other families of insects are attracted to the body solely because they feed on the necrophagous insects that are present.

1.1 History of forensic entomology

Insects are known to have been used in the detection of crimes for a long time and a number of researchers have written about the history of forensic entomology (Benecke, 2001; Greenberg and Kunich, 2002). The Chinese used the presence of flies and other insects as part of their investigative armoury for crime scene investigation and instances of their use are recorded as early as the mid-tenth century (Cheng, 1890; cited in Greenberg and Kunich, 2002). Indeed, such was the importance of insects in crime scene investigation that in 1235, a training manual on investigating death, *Washing Away of Wrongs*, was written by Sung Tz'u. In this medico-legal book it is recorded that the landing of a number of blowflies on a particular sickle caused a murderer to confess to murdering a fellow Chinese farm worker with that sickle.

Between the thirteenth and the nineteenth century, a number of developments in biology laid the foundation for forensic entomology to become a branch of scientific study. The two most notable were, perhaps, experiments in Italy by Redi (1668) using the flesh of a number of different animal species, in which he demonstrated that larvae developed from eggs laid by flies, and the work by Linnaeus (1775) in developing a system of classification. In so doing, Linnaeus provided a means of insect identification, including identifying such forensically important flies as *Calliphora vomitoria* (Linnaeus). These developments formed foundations from which determination of the length of the stages in the insect's life cycle could be worked out and indicators of time since death could be developed.

A particularly significant legal case, which helped establish forensic entomology as a recognized tool for investigating crime scenes, was that of a murdered newborn baby. The baby's mummified body, encased in a chimney, was revealed behind a mantelpiece in a boarding house when renovation work was being undertaken in 1850. Dr Marcel Bergeret carried out an autopsy on the body and discovered larvae of a fleshfly, *Sarcophaga carnaria* (Linnaeus), and some moths. He concluded that the baby's body had been sealed up in 1848 and that the moths had gained access in 1849. As a result of this estimation of post mortem interval, occupiers of the house previous to 1848 were accused and the current occupiers exonerated (Bergeret, 1855).

The next significant point in the history of forensic entomology resulted from observations and conclusions made by Mégnin (1894). He related eight stages of human decomposition to the succession of insects colonizing the body after death. He published his findings in *La Faune des Cadavres: Application de l'Entomologie à la Médicine Légale*. These stages of decomposition were subsequently shown to vary in speed and to be dependent upon environmental conditions, including temperature and, for example, whether or not the corpse was clothed. However, the similarity in overall decomposition sequence and the value of the association of insects has been demonstrated for decomposition of the bodies of a number of animal species. This knowledge about insect succession on a corpse became the basis for forensic entomologists' estimations of the time since death of the corpse.

In the twentieth century insects were shown to be of value in court cases involving insect colonization of body parts recovered from water and not just whole corpses found on land. On 29 September 1935, several body parts, later identified as originating from two females, were recovered from a Scottish river near Edinburgh. The identities of the deceased were determined and the women were named as Mrs Ruxton and Mary Rogerson, 'nanny' for the family. The presence of larvae of the blowfly *Calliphora vicina* Robineau-Desvoidy, in their third larval instar, indicated that the eggs had been laid prior to the bodies being dumped in the river. This information, combined with other evidence, resulted in the husband, Dr Ruxton, being convicted of the murder of his wife and Mary Rogerson.

The acceptance of forensic entomology has depended upon both academics and practitioners working alongside the police and legal authorities, to refine and develop forensic entomology as a scientific study in the late twentieth and early twenty-first centuries. A list of forensic entomologists, who are members of the American Board of Forensic Entomology, the European Association for Forensic Entomology and other professional entomological and medical organizations, can be found on the website: http://folk.uio.no/mostarke/forensic_entomologists. html

1.2   Indicators of time of death

In the first 72 hours after death, the pathologist is usually considered to be able to provide a reasonably accurate determination of the time of death. Historically, this has been based upon the condition of the body itself and such features as the fall in body temperature. Beyond this time, there is less medical information with which to correlate post mortem interval (PMI). So another area of expertise is required to clarify time of death. The forensic entomologist can provide a measure of the possible post mortem interval, based upon the life cycle stages of particular fly species recovered from the corpse, or from the succession of insects present on the body. This estimate can be given over a period of hours, weeks or years. The start of the post mortem interval is considered to coincide with the point when the fly first laid its eggs on the body, and its end to be the discovery of the body and the recognition of life stage of the oldest colonizing species infesting it. The duration of this stage, in relation to the particular stage of decay, gives an accurate measure of the probable length of time the person has been dead and may be the best estimate that is available.

1.3   Stages of decomposition of a body

The stages of decomposition of a body have been a topic of interest for both artists and scientists over a long period of time (Figures 1.1–1.8). There are three recognizable processes in corpse decomposition. These are autolysis, putrefaction and skeletal bone decomposition (diagenesis). In autolysis, a process of natural breakdown, the cells of the body are digested by enzymes, including lipases, proteases and carbohydrases. This process can be most rapid in organs such as the brain and liver (Vass, 2001). A 'soup' of nutrients is released which forms a food source for bacteria. Putrefaction is the breakdown of tissues by bacteria. As a result, gases such as hydrogen sulphide, sulphur dioxide, carbon dioxide, methane, ammonia, hydrogen and carbon dioxide are released. Alongside this, anaerobic fermentation takes place when the volatiles propionic and butyric acid are formed. The body undergoes active decay, in which protein sources are broken

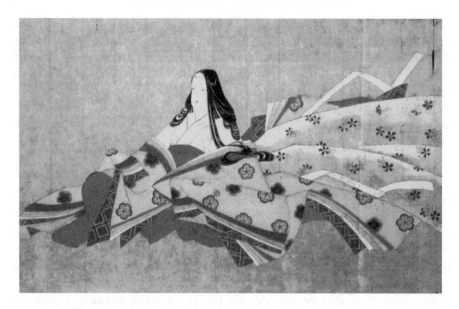

Figure 1.1 Artistic impressions of stages of decomposition of the body (Morishige, 1673–1680). The nine contemplations of the impurity of the human body, stage 1–9. Reproduced with permission of the Etnografisch Museum, Antwerp, Belgium inv.nrs: AE 4552 1/20–19/20 (A colour reproduction of this figure can be found in the colour section towards the centre of the book)

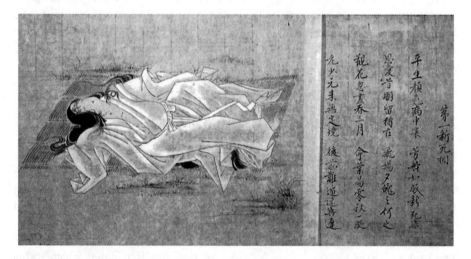

Figure 1.2 Artistic impressions of stages of decomposition of the body (Morishige, 1673–1680). The nine contemplations of the impurity of the human body, stage 1–9. Reproduced with permission of the Etnografisch Museum, Antwerp, Belgium inv.nrs: AE 4552 1/20–19/20 (A colour reproduction of this figure can be found in the colour section towards the centre of the book)

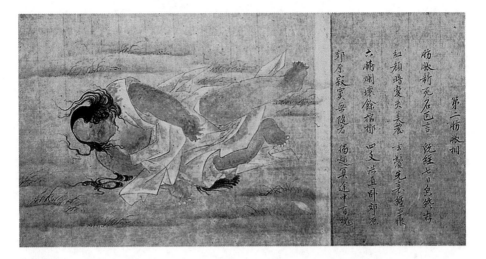

Figure 1.3 Artistic impressions of stages of decomposition of the body (Morishige, 1673–1680). The nine contemplations of the impurity of the human body, stage 1–9. Reproduced with permission of the Etnografisch Museum, Antwerp, Belgium inv.nrs: AE 4552 1/20–19/20 (A colour reproduction of this figure can be found in the colour section towards the centre of the book)

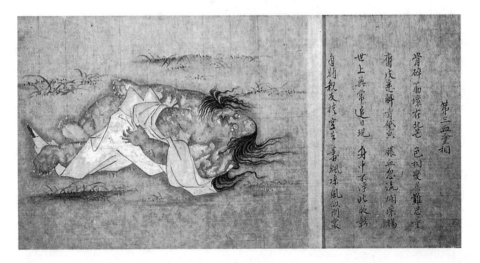

Figure 1.4 Artistic impressions of stages of decomposition of the body (Morishige, 1673–1680). The nine contemplations of the impurity of the human body, stage 1–9. Reproduced with permission of the Etnografisch Museum, Antwerp, Belgium inv.nrs: AE 4552 1/20–19/20 (A colour reproduction of this figure can be found in the colour section towards the centre of the book)

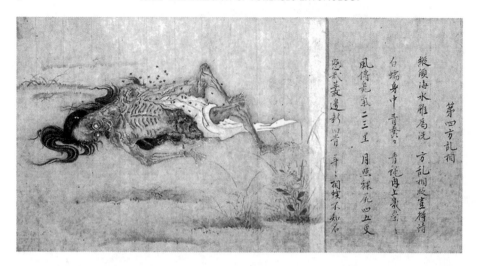

Figure 1.5 Artistic impressions of stages of decomposition of the body (Morishige, 1673–1680). The nine contemplations of the impurity of the human body, stage 1–9. Reproduced with permission of the Etnografisch Museum, Antwerp, Belgium inv.nrs: AE 4552 1/20–19/20 (A colour reproduction of this figure can be found in the colour section towards the centre of the book)

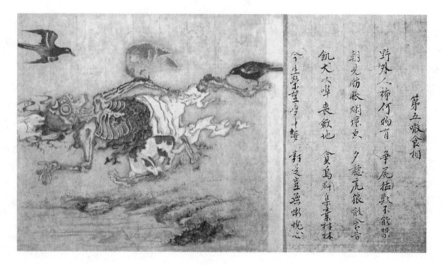

Figure 1.6 Artistic impressions of stages of decomposition of the body (Morishige, 1673–1680). The nine contemplations of the impurity of the human body, stage 1–9. Reproduced with permission of the Etnografisch Museum, Antwerp, Belgium inv.nrs: AE 4552 1/20–19/20 (A colour reproduction of this figure can be found in the colour section towards the centre of the book)

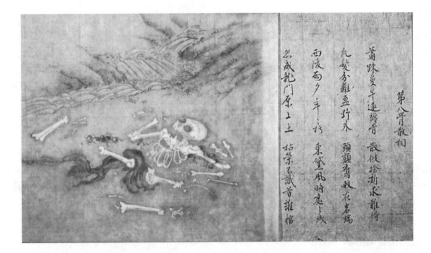

Figure 1.7 Artistic impressions of stages of decomposition of the body (Morishige, 1673–1680). The nine contemplations of the impurity of the human body, stage 1–9. Reproduced with permission of the Etnografisch Museum, Antwerp, Belgium inv.nrs: AE 4552 1/20–19/20 (A colour reproduction of this figure can be found in the colour section towards the centre of the book)

Figure 1.8 Last stage of decomposition of a human body (A colour reproduction of this figure can be found in the colour section towards the centre of the book)

down into fatty acids by bacteria (Vass, 2001). Fatty acids and such compounds as skatole, putrescine and cadaverine are significant members of these decomposition products (although Vass *et al.*, 2004, commented on their absence from recovered volatiles from buried bodies).

When the soft tissue is removed, skeletal material – organic and inorganic remains – are further broken down by environmental conditions and are finally reduced to components of the soil. The rate of decomposition is temperature-dependent. A formula has been proposed by forensic pathologists to estimate the time of body decomposition to a skeleton, in relation to temperature (Vass, 2001). The formula is:

$$Y = 1285/X$$

where Y is the number of days to mummification, or skeletonization, and X is the average temperature for the days before the body was found (Vass $et al.$, 1992).

1.3.1 On land

The body can be allocated to one of five recognizable post mortem conditions, which can be linked to the eight waves of arthropod colonization proposed by Mégnin (1894). No distinction from one stage to the next is obvious and Gaudry (2002), on the basis of 400 cases, considers Mégnin's first two waves to be one. Although no stage has a fixed duration, each stage can be associated with a particular assemblage of insects. The profiles of insects would appear to be universal, although the majority of research on this aspect has, until recently, been undertaken in North America (Hough, 1897; Easton and Smith, 1970; Rodriguez and Bass, 1983; Catts and Haskell, 1990; Mann, Bass and Meadows, 1990; Goff, 1993; Dillon and Anderson, 1996; VanLaerhoven and Anderson, 1999; Byrd and Castner, 2001). These stages of post mortem change are:

- *Stage 1: Fresh stage.* This stage starts from the moment of death to the first signs of bloating of the body. The first organisms to arrive are the blowflies (the Calliphoridae). In Britain these are usually *Calliphora vicina* or *Calliphora vomitoria* Linnaeus, or in early spring may be *Protophormia* (= *Phormia*) *terraenovae* Robineau-Desvoidy) (Nuorteva, 1987; Erzinçlioğlu, 1996).

- *Stage 2: Bloated Stage.* Breakdown of the body continues because of bacterial activity, or putrefaction, and this is perhaps the easiest stage to distinguish. Gases causing the corpse to bloat are generated through metabolism of nutrients by anaerobic bacteria. Initially the abdomen swells but later the whole body becomes stretched like an air-balloon (Figure 1.9). At this stage more and more blowflies are attracted to the body, possibly in response to the smell of the breakdown gases. Vass $et al.$ (1992, 2004) studied the odours emanating from dead bodies that were both resting on the surface and had been buried. Their work provides clarification of the identity of some of these gases and the information supplements that provided by Mégnin (1894); Hough (1897) and Smith (1986).

  Rove beetles (Staphylinidae) may be attracted to the body at the bloat stage because of the 'ready meals' of eggs and maggots. These and other predators

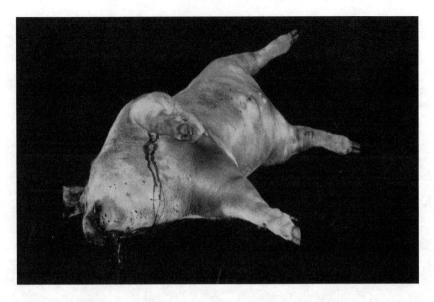

Figure 1.9 Body in bloat (A colour reproduction of this figure can be found in the colour section towards the centre of the book)

can affect the interpretation of the range of insects and insect life stages present as they feed on larvae or remove puparia (Smith, 1986).

- *Stage 3: Active decay stage.* This stage is recognizable by the skin of the corpse breaking up and starting to slough from the body. This sloughing allows the decomposition gases to escape and so the inflation of the body gradually subsides as putrefaction continues. In the later stages of putrefaction fermentation occurs and butyric and caseic acids are generated. This is followed by a period of advanced putrefaction, which includes ammoniacal fermentation of the body, to which a different cohort of insects are attracted. These include the silphid beetle *Nicrophorus humator* (Gleditsch) the histerids *Hister cadaverinus* (Hoffmann) and *Saprinus rotundatus* Kugelann, and the muscid fly *Hydrotaea capensis* Wiedeman (= *Ophyra cadaverina* Curtis).

- *Stage 4: Post-decay stage.* In the later stages of decay, all that remains of the body are skin, cartilage and bones with some remnants of flesh including the intestines. Any remaining body tissue can be dried. The biggest indicator of this stage is an increase in the presence of beetles and a reduction in the dominance of the flies (Diptera) on the body.

- *Stage 5: Skeletonization.* At this stage the body is only hair and bones (Figure 1.10). No obvious groups of insects are associated with this stage, although beetles of the family Nitidulidae can, in some instances, be found. The body has clearly reached its final stage of decomposition. Any further breakdown is best described in terms of the decay of individual components of the body, such as the bones of the feet and legs, the skull and the ribs.

Figure 1.10 Post-decay stage of human decomposition. The breakdown material was retained within the polythene

1.3.2 Submerged in water

In water these same five stages still occur along with an additional stage. This additional stage is the floating decay stage, where the body rises to the water surface. At this point, besides aquatic insects such as midge (chironomid) larvae and invertebrates such as water snails, terrestrial insect species also colonize the body.

This stage is the most obvious stage and tends to be the point at which a body is noticed and recovered from the water. The period of time after death when this takes place will depend on the temperature of the water.

The relationship between time of death and physical breakdown of the body has been investigated by Giertsen (1977). He cited Casper's Dictum as a means of determining the length of the post mortem interval. This rule says that:

> '. . . at a tolerable similar average temperature, the degree of putrefaction present in a body lying in the open air for one week (month) corresponds to that found in a body after lying in the water for two weeks (months), or lying in the earth in the usual manner for eight weeks (months)'.

The reason for this difference in decomposition is that the speed at which the body loses heat in water is twice the speed at which the body will lose heat in air.

Box 1.1   Hint

Besides skeletonization, with the resultant change in the bone structure (diagenesis), two other outcomes of the decomposition process may occur. These are mummification and the generation of 'grave wax' or adipocere.

Diagenesis

When the body reaches the skeletal stage, changes to the bone called diagenesis occur. Diagenesis is defined, in chemical terms, according to *Collins Dictionary of the English Language* (Hanks, 1984), as recrystallization of a solid to form large crystal grains from smaller ones. The changes in the bone structure depend upon the breakdown of the soft tissue. This is affected by the nature of the death and subsequent treatment of the body, including the type of environment in which the body is buried.

Investigating bone can tell us about the latter stages of post mortem change because a number of features can be quantified. The amount of post mortem change can be estimated if the bone histology is investigated under the microscope, the degree of bone porosity is determined; the carbonate and protein content of the bone are calculated; the crystalline nature and content of the bone mineral made of calcium fluorophosphate or calcium chlorophosphate (apatite) is examined, and which components have leached out of or into the bone are determined.

Insect attack, both before and after the body is buried, has a role to play in causing change to the environment and hence bone diagenesis.

Adipocere

If the body is in an environment which combines a high humidity with high temperatures, the subcutaneous body fat of the face, buttocks (breasts in the female) and the extremities become hydrolysed. Fatty acids are released. These form food for bacteria, which can speed up the rate at which adipocere is made. For example, *Clostridium* bacteria will convert oleic acid (a fatty acid) into hydroxystearic acid and oxostearic acid.

Two types of adipocere are found, depending on whether the fatty acids combine with sodium or with potassium. If sodium from the breakdown of intercellular fluid is bound to the fatty acids, the adipocere is hard and curly. Where the cell membranes break down and potassium is released, a softer adipocere results, which is often termed 'pasty'. An indication of submergence in cold water is the uniform cover of adipocere over the body (Spitz, 1993).

Mummification

If water is removed from skin and tissue, that tissue becomes desiccated and mummification will occur. This happens particularly where a body is kept in an environment with a dry heat, little humidity and where the airflow is good. Chimneys are examples of good locations with these features. In mummified corpses in temperate conditions, the extremities become shrivelled and the skin tends to be firm but wrinkled and to have a brown colouration. The internal organs, such as the brain, will decompose during mummification.

More research is needed to explore decomposition in various types of water body and in a number of locations, so that a comprehensive picture of the potential indicators of submerged post mortem interval can be clarified. Research by Keiper and Casamatta (2001), Hobischak and Anderson (2002) and Merritt and Wallace (2001) has provided a starting point.

Whilst the major contribution of forensic entomology to solving crimes could be considered to be in relation to suspicious death, it has a part to play in investigating other crimes in which the victims may be alive or dead.

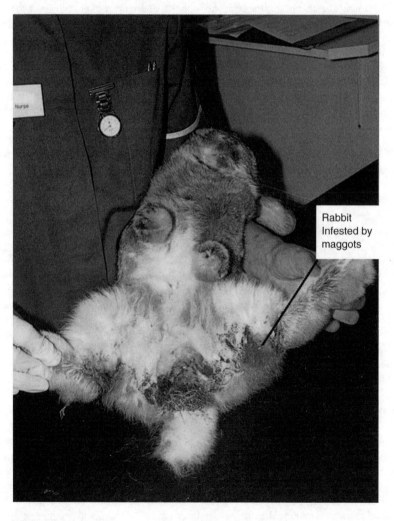

Figure 1.11 Rabbit exhibiting myiasis. Reproduced with kind permission of Frances Harcourt-Brown

1.4 Indicators of physical abuse

Insects are of value as forensic indicators in cases of neglect or abuse. Some insects, for example the greenbottle *Lucilia sericata* (Meigen), are attracted to odours, such as ammonia, resulting from urine or faecal contamination. Adult flies of this species tend to be attracted to an incontinent individual; a baby that has not had its nappy changed sufficiently often, or incontinent old people who have not been assisted in maintaining their bodily hygiene.

Flies may lay their eggs in clothing or on skin. These eggs, if undiscovered, will hatch into maggots (larvae) which start feeding upon flesh, or on wounds, ulcers or natural entry points of the body. Over time the flesh will be eaten away and the region may be further infected by bacteria as well as being invaded by other insects.

Such an insect attack can also happen to animals. In particular, rabbits, pigs, dogs and sheep can be the victims of fly strike (Figure 1.11) because of urine or faecal material attached to their fur, fleece or hind quarters through neglect, poor caging and living conditions or ill-health reflected by 'scouring'. Such cases are considered to be instances of physical abuse, since victims are unable to remove the eggs or maggots themselves. The results can be serious, requiring attention

---

Box 1.2   Hint

The invasion of living tissue by insects is also of concern to the forensic entomologist. This invasion is called *myiasis* and becomes relevant where cases of misuse and abuse are involved.

Myiasis has been defined according to two criteria: the biological requirements of the fly, or where the flies attack the organism, be it human or animal. James (1947) defined biological myiasis as invasion of tissue or organs of man or animals by dipterous larvae. He acknowledged Patton's (1922) earlier views that the presence of eggs, pupae or adults might be included, but considered that the larval stage was the 'active stage' of myiasis.

In medical terms, myiasis can be defined according to the location of the fly infestation. For example, it can be defined as: wound myiasis; myiasis of the nose, mouth and accessory sinuses; aural myiasis; ocular internal and external myiasis; myiasis of the rectal region and vagina; myiasis of the bladder and urinary passages; furuncular, dermal and sub-dermal myiasis; creeping dermal, sub-dermal myiasis or enteric myiasis.

Flies such as *Lucilia sericata*, *Musca domestica* Linnaeus and *Phormia regina* Meigen, the initial colonizers of the body, have all been implicated in cases of myiasis.

from veterinary surgeons and even leading to the death of the animal, or requiring its euthanasia.

Care however, has to be taken in making assumptions about the existence of physical abuse or assault prior to death. Work by Komar and Beattie (1998) in studies on dressed pigs, showed the effect of bloat was to cause the same disturbance and tearing of clothes which are characteristic of sexual assault. They considered that maggot masses were particularly important in deriving such changes to clothing.

1.5   Insect larvae: a resource for investigating drug consumption

The insect life cycle stage that feeds on the cadaver is a potential reservoir of undigested flesh from the corpse. Because, in some circumstances, the flesh from the corpse can retain some types of drugs that had been consumed by the victim before he/she died and which may even have been the cause of death, these drugs may be recovered by analysing the insects and may include opiates (Introna *et al.*, 1990), the barbiturate phenobarbital, benzodiazepines or their metabolites, such as oxazepam, triazolam, antihistamines, alimemazine and chlorimipramine, a tricyclic antidepressant (Kintz *et al.*, 1990; Sadler *et al.*, 1995).

To date there is not a great deal of information available that indicates the role of drugs, which are present in decomposing body tissue, on necrophagous larvae. Musvaska *et al.* (2001) examined the effects of consuming liver containing either a barbiturate (sodium methohexital) or a steroid (hydrocortisone) on the development of a fleshfly, *Sarcophaga* (= *Curranea*) *tibialis* Macquart. They showed that, compared with controls, the length of the larval stage was increased, whilst pupariation was more rapid. In laboratory experiments investigating the effects of heroin, Arnaldos *et al.* (2005) also showed that the length of time taken to complete individual larval stages in *Sarcophaga tibialis* was considerably longer, in contrast to those larvae which were not fed heroin.

However, heroin has been shown to increase the rate at which other species of maggots (e.g. *Boettcherisca peregrina* Robineau-Desvoidy) grow, whilst increasing the duration of pupal development (Goff *et al.*, 1991). Cocaine and one of its breakdown products has been found in small amounts in the puparium of Calliphoridae (Nolte *et al.*, 1992), so this drug is clearly sequestered in the larval body and retained in the next life stage. However, Hédouin *et al.* (2001) only showed a correlation between concentration of morphine in body tissues and that in the tissue of larvae of *Protophormia terraenovae* and *Calliphora vicina* in the third instar. In *Lucilia sericata* they found that the post mortem interval could, in reality, be 24 hours longer than expected (Bourel *et al.*, 1999).

Suicide can be investigated using forensic entomology. By analysing the maggots which had fed on the corpse and demonstrating the presence in the body of malathion, an organophosphate insecticide, Gunatilake and Goff (1989) confirmed that a 58 year-old man had committed suicide. A bottle of malathion had been found near to the corpse.

Miller *et al.* (1994) analysed chitinized insect tissue from mummified remains from which the normal toxicological sources were absent. They were able to recover amitriptyline and nortryptyline from the puparia of scuttle flies (Phoridae) and the exuviae of hide beetles (Dermestidae). Sadler *et al.* (1997), however, found that there was a variation in larval drug accumulation of amitriptyline and urged caution in directly relating the concentration harvested from the larvae to concentrations in the original source.

Paying attention to the facts known about the lifestyle of the victim may assist in interpreting the post mortem interval, using the developmental stage of the insect recovered from the body. So, all of the information known about the crime scene and pre mortem behaviour of the person should be taken into account when investigating the entomological evidence.

Insects collected with plant material destined to be used illegally can indicate the part of the world from which the plants originated. This information may be of forensic value to Customs and Excise Officers. For example, in two separate drug seizures in New Zealand, cannabis was apprehended along with eight Asian species of beetles, as well as wasps and ants. The beetles were identified by Dr Trevor Crosby as belonging to the families Carabidae, Bruchidae and Tenebrionidae. By looking at the geographic distribution of all of the insects and the level of overlap of their distributions, entomologists concluded that the cannabis came from the Tenasserim region, between the Andaman Sea and Thailand. One of two suspects confessed on the basis of this evidence (Crosby *et al.*, 1986).

1.6 Insect contamination of food

Many societies consume insects as part of their diet (Figure 1.12). For example, aquatic beetles such as the giant water bug, *Lethocerus indicus* Lepeletier Serville, are eaten as a delicacy across south-eastern Asia. Chocolate-covered bees have been sold in speciality shops in the UK, and in North America some shops sell canned, fried grasshoppers (DeFoliart, 1988; Menzel and D'Aluisio, 1998), whilst Thai cooked crickets in tins are available via the world-wide web.

However, the presence in food of insects that are eaten unintentionally, or could be eaten along with the food, is considered unacceptable to the consumer and a source of contamination. For example, the saw-toothed grain beetle, a stored product pest, may be found in cereal packages; wire worms may be sold along with freshly cut lettuces, or may be processed into lettuce and tomato sandwiches; whilst in many countries, fish and meat which is left in the open to dry can become infested with beetles or flies, either in the drying process or later on a market stall. These are then eaten and have the potential to cause illness. Forensic entomologists may therefore find themselves being asked for an expert opinion in civil cases

Figure 1.12   A tequila bottle label illustrating the Maguey worm, *Aegiale hesperiatis* Walker (Lepidoptera), which authenticates the drink. Reproduced from a letter in *Antenna* **6**(3) (1982) with kind permission of Dr B. Lawrence and the Royal Entomological Society of London

relating to the food industry, where food has been contaminated by insects living in close association with man (such insects are described as 'synanthropic').

1.7 Further reading

Amendt J., Krettek R. and Zehner R. 2004. Forensic entomology. *Naturwissenschaften* **91**: 51–65.

Anderson G. S. 1999. Wildlife forensic entomology: determining time of death in two illegally killed black bear cubs. *Journal of Forensic Sciences* **44**(4): 856–859.

Benecke M. 2001. A brief history of forensic entomology. *Forensic Science International* **120**: 2–14.

Benecke M. 2004. Arthropods and corpses. In Tsokos M. (ed.), *Forensic Pathology Reviews 2*. Humana: Totowa, NJ; pp 207–240.

Byrd J. H. and Castner J. L. 2001. *Forensic Entomology: The Utility of Arthropods in Legal Investigations*. CRC Press: Boca Raton, FL.

Catts E. P. and Haskell N. H. (eds). 1990. *Entomology and Death: A Procedural Guide*, Joyce's Print Shop: Clemson, SC.

Catts E. P. and Goff M. L. 1992. Forensic entomology in criminal investigations. *Annual Review of Entomology* **37**: 253–272.

Erzinçlioğlu Y. Z. 2000. *Men, Murder and Maggots*. Harley Press: Colchester.

Goff M. L. 1993. Estimation of post mortem interval using arthropod development and successional patterns. *Forensic Science Review* **5**: 81–94.

Goff M. L. 2000. *A Fly for the Prosecution*. Harvard University Press: Cambridge, MA.

Goff M. L. and Lord W. D. 1994. Entomotoxicology: a new area for forensic investigation. *American Journal of Forensic Medicine and Pathology* **15**(1): 51–57.

Greenberg B. and Kunich J. C. 2002. *Entomology and the Law*. Cambridge University Press: Cambridge.

Gupta A. and Setia P. 2004. Forensic entomology – past, present and future. *Aggrawal's Internet Journal of Forensic Medicine and Toxicology* **5**(1): 50–53.

Haefner J. M., Wallace J. R. and Merritt R. W. 2004. Pig decomposition in lotic aquatic systems: the potential use of algal growth in establishing a post mortem submersion interval (PMSI). *Journal of Forensic Sciences* **49**(2): 1–7.

Haskell N. H., McShaffrey D. G., Hawley D. A., Williams R. E. and Pless J. E. 1989. Use of aquatic insects in determining submersion times. *Journal of Forensic Sciences* **34**(3): 623–632.

Hawley D. A., Haskell N. H., McShaffrey D. G., Williams R. E. and Pless J. E. 1989. Identification of a red 'fiber': chironomid larvae. *Journal of Forensic Sciences* **34**(3): 617–621.

Latham P. 1999. Edible caterpillars of the Bas Congo region of the Democratic Republic of the Congo. *Antenna* **23**(3): 134–139.

O'Brien C. and Turner B. 2004. Impact of paracetamol on *Calliphora vicina* larval development. *International Journal of Legal Medicine* **118**(4): 188–189.

Sachs Snyder J. 2002. *Time of Death: The True Story of the Search for Death's Stopwatch*. QPD: London.

Vass A. A., Bass W. B., Wolt J. D., Foss J. E. and Ammons J. T. 1992. Time since death determinations of human cadavers using soil solution. *Journal of Forensic Sciences* **37**(5): 1236–1253.

Vass A. A., Smith R. R., Thompson C. V., Burnett M. S. *et al.* 2004. Decompositional odour analysis database, *Journal of Forensic Sciences* **49**(4): 1–10.

Wood M., Laloup M., Pien K., Samyn N. *et al.* 2003. Development of a rapid and sensitive method for the quantitation of benzodiazepines in *Calliphora vicina* larvae and puparia by LC–MS–MS. *Journal of Analytical Toxicology* **27**(7): 505–512.

Useful websites

Pounder D. J. 1995. Postmortem changes and time of death: www.dundee.ac. uk / forensicmedicine/IIb/timedeath.htm

Morten Staerkeby website: www.folk.uio.no/mostarke/forensic_ent/forensic_ entomology.htm

European Association for Forensic Entomology: http://new.eafe.org

American Board of Forensic Entomology: www.missouri.edu/~agwww/entomology

2

Identifying flies that are important in forensic entomology

In order to interpret a crime scene, it is important to know which species of insects are infesting the body and something about their habits and environmental requirements. To make these identifications you will need to know what the parts of the insect are called, so that you can use the scientific keys written by taxonomists or identify the species of insect by some other scientific means, such as DNA analysis, which would be acceptable to a court.

Insects are recognizable because they have a hardened body case (an exoskeleton) which is split into three. These sections are called the *head*, the *thorax* and the *abdomen*. The sections have three dimensions, a top (*dorsum*), an underneath (*sternum*) and sides, each of which is a called a *pleuron*. The thorax is split into three segments which may or may not be clearly defined, depending upon the species. Starting from the head and working backwards along the thorax, these segments are called the *prothorax*, the *mesothorax* and the *metathorax*.

All insects have six legs (three pairs of jointed legs). These too are made up of sections; starting at the point nearest to the body (the proximal region), we have sections called the *coxa* (plural coxae), followed by the *trochanter* (a small section), the *femur*, *tibia* and the *tarsus* (plural tarsi) (Figure 2.1). The numbers of tarsal segments (*tarsomeres*) may vary, but usually there are five per leg. The legs are located on the thorax. One pair of legs is found on each of the thoracic segments, i.e. one pair is on the prothorax, the second pair is on the mesothorax, whilst the third pair of legs is attached at the metathorax.

The first pair of the insect's membranous wings is attached to the mesothorax. A second pair of wings (or a modification of the second pair) is attached at the metathorax – the final segment of the thorax. The membranous wings are supported by veins, which have been named in order to assist in identifying the species of insect. They are counted from the first vein running along the wing edge (the system of naming described here is based on work by two scientists called Comstock and Needham; in consequence it is called the Needham–Comstock system).

The first of the wing veins, *vein 1*, is called the *costa*. This is a thick, hardened vein and gives the wing some rigidity for flying. The second vein, *vein 2*, is called the stem vein or *subcosta*. The third vein at the proximal (body) point

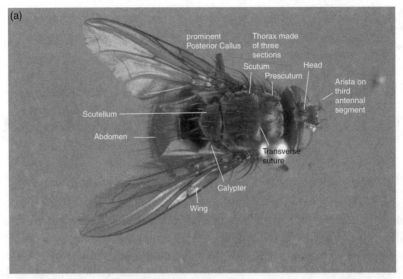

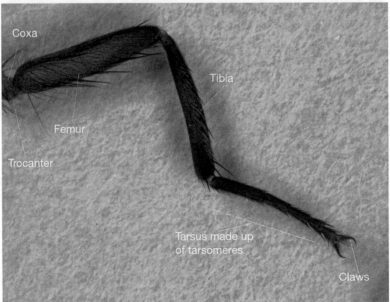

Figure 2.1 Structure of (a) the insect and (b) the leg

of attachment, *vein 3*, is called the *radius*. The fourth long vein, *vein 4*, is the *media* (or medial vein). This can be split into four veins as it passes to the wing edge. The fifth vein, *vein 5*, is called the *cubitus* vein and in some insect species also splits. In addition, there are several *cross veins* and the rear surface of the wing can be modified in order to give a surface area which is flexible, to assist in flight and movement whilst in the air. These wing cross veins are named on

Box 2.1   Hint The system of naming bristles on a fly

How to name leg bristles

Bristles can be distributed down the femur at set points. A bristle near the joint of the femur with the trocanter is called a basal bristle. The bristle near the joint with the tibia is called an apical bristle. The bristle just before the joint with the tibia is the pre-apical bristle. The one just before the joint with the coxa is a sub-basal bristle.

Last century, a system was proposed which helps us to decide the names of the other bristles (Grimshaw, 1917, 1934). The method requires that you think of the bristle position in terms of a cross-section through the leg as if it is stretched out horizontally from the body (Figure 2.1a indicates the positions of bristles when the leg is held horizontally). The positions of the bristles are thought of as positions on a clock face: at 12 o'clock the bristle would be called a *dorsal* bristle; at 6 o'clock a *ventral* bristle; at 10 o'clock an *antero-dorsal* bristle; and at 2 o'clock a *postero-dorsal* bristle.

Moving round the clock face, at the 3 o'clock position, as the leg is held horizontally from the body, the bristle is called a *posterior* bristle; at the 9 o'clock position an *anterior* bristle; at the 7 o'clock position an *antero-ventral* bristle; and at the 5 o'clock position a *postero-ventral* bristle. In identification keys, having a bristle present on any of the three pairs of legs constitutes presence of a bristle. It does not necessarily need to be present on all three pairs of legs.

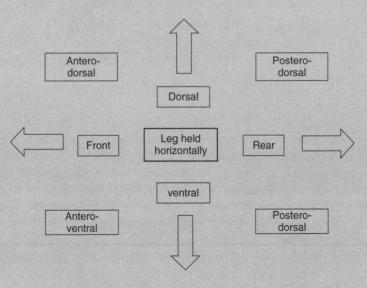

Figure 2.1a   Orientations of bristles on the insect leg

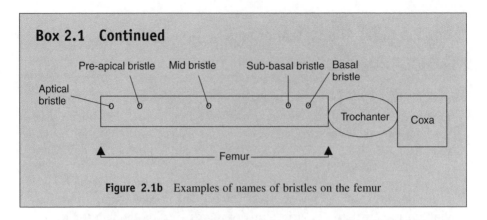

Figure 2.1b Examples of names of bristles on the femur

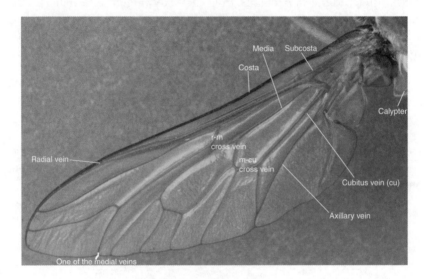

Figure 2.2 The structure of the insect wing

the basis of the veins between which they pass. The *c* and *sc* vein runs between the costa and subcosta. The *r* veins run between the splits in the radial vein. The *r-m* vein runs between the radial and the median veins and the *m-cu* veins run between the media and the cubitus veins. An example of the structure of the wing is shown in Figure 2.2. The vein numbers are hard to work out from first principles and diagrams for the group should always be consulted. One book may not be consistent with another.

In addition, insects have a pair of segmented structures, positioned antero-dorsally on the head, which are sense organs; these are the *antennae*, which are commonly called 'feelers'. There are many different forms of antennae and their shapes can assist in identification (Figure 2.3). The antennae provide the insect

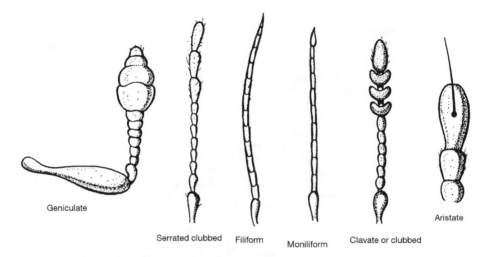

Geniculate

Serrated clubbed   Filiform   Moniliform   Clavate or clubbed   Aristate

Figure 2.3 Examples of some of the forms of antennae found in insects. Reproduced from Munro (1966) with kind permission of Rentokil Initial plc

with a means of gaining both chemical (contact chemoreceptors) and mechanical information (mechanoreceptors over a distance) from its surroundings

Insects (a *class* within the *phylum* Arthropoda, or jointed limbs phylum) are divided into a large number of groups called *orders*. Each order is divided into a number of families. Each *family* is made up of a number of *genera* (singular *genus*) and each genus has one or more *species* (Figure 2.4). The named groups, at each level of this hierarchy, are called *taxa* (singular *taxon*).

One of the orders of insects which are forensically relevant is the Order Diptera – the true or two-winged flies.

2.1   What is a fly and how do I spot one?

Flies are easily distinguished from other insects by having two fully developed, usually obvious, front wings, but with each of its two back wings modified into balancers, called *halteres*. These structures resemble tiny drumsticks.

There has been a considerable change in fly taxonomy recently and the agreed groupings of flies arise from modern developments in taxonomy, including molecular studies. In the older forensic text books, the taxonomy by Kloet and Hincks (1976) has been used. In this classification, the Diptera were divided into three suborders, with the third, the Cyclorrhapha, subdivided into the Aschiza, Schizophora-Acalyptratae and the Schizophora-Calyptratae.

Currently, the phylogenetic classifications between the *suborders* and the families are mainly a response to practical considerations, so that now, rather than three suborders within the Diptera, there are two suborders

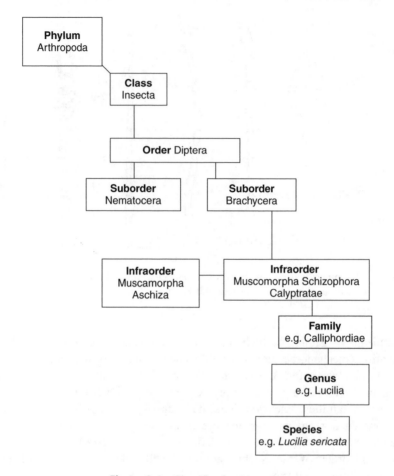

Figure 2.4 Classification hierarchy

The first of these suborders, the *Nematocera*, or thread horns, contains the crane flies (see Figure 2.5). These insects have a long slender body, long antennae with more than six segments and a complex wing venation. The Nematocera have mouthparts which are 'droopy' (pendulous). The larvae of this suborder have a structurally distinct (exserted) head with horizontal biting mouthparts. The pupa is not encased and so morphological structures such as wing buds are visible. The winter gnats (Trichoceridae) are an example of a nematoceran family which has members that are forensically relevant. They have been used to determine post mortem interval in the winter, when many other insects are no longer available.

The second suborder is called the *Brachycera*, or short horns. These are much more robust flies and are often called 'the higher flies' (Figure 2.6). They have shorter antennae than the Nematocera, with eight or fewer segments. Their wing venation is less complex and their pupae grow inside a case which is the hardened remains of the last larval (third instar) coat. The male genitalia are separated into

Figure 2.5 (a) A nematoceran fly, a member of the Tipulidae; note the halteres, or balancers, and the pendulous mouthparts. (b) Nematoceran wing showing the complexity of wing venation (A colour reproduction of this figure can be found in the colour section towards the centre of the book)

two parts. The larvae have an elongated head which is combined within the first segment of the thorax (prothorax). Their mandibles are divided.

The Brachycera are split into groups (*infraorders*), called the Tabanomorpha and Asilomorpha, which also made up the Brachycera under the old classification system. A third group (infraorder), the Muscomorpha, is most important forensically. These are predominantly the Cyclorrhapha of the old classification

Figure 2.6   An example of a member of the Brachycera; a tabanid or horse fly

system. This group have *antennae* with a bristle and have three larval stages in which the morphological distinction into head and body of the larva is absent.

This group is subdivided into:

1. *The Muscomorpha Aschiza*. In this division the depression and suture over the antenna is either absent or very indistinct. Some wing veins close off a wing compartment ('cell') called the *anal* cell (Figure 2.7). This is very long and closes, or almost closes, at a point at least more than half to two-thirds of the way to the edge of the wing. The Phoridae are of forensic note in this division.

2. Usually, however, the anal wing cell in insects of forensic importance is short or even not there at all. This is a feature of *the Muscomorpha Schizophora*, which includes the blowflies, cheese skippers and the fleshflies. This group is divided into two, depending upon whether or not there are flaps or *calypters* (Figure 2.8) at the base of the wing, with the lower one joining the wing to the thorax.

3. *The Muscomorpha Schizophora Acalyptratae*. These flies emerge from the puparial case by using an 'air bag' or *ptilinum*. This deflates and draws back below the eyes and above the antennae. Its presence is indicated by a depression like a crease or furrow, just below the eyes, called a *ptilinal suture* (Figure 2.9). Flies in this group do not have a thorax with sutures that completely divide the prothorax. The halteres are exposed and the antennae do not have a slit in segment three. Such flies are termed *acalypterate*. Of forensic note are the Piophilidae, Sphaeroceridae and Sepsidae.

Figure 2.7   The wing of a member of the infraorder Muscomorpha Aschiza, illustrating the long anal cell

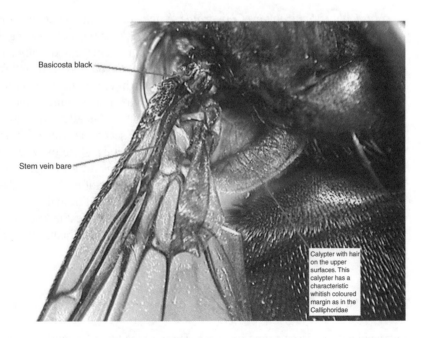

Basicosta black

Stem vein bare

Calypter with hair on the upper surfaces. This calypter has a characteristic whitish coloured margin as in the Calliphoridae

Figure 2.8   The wing calypter (squama)

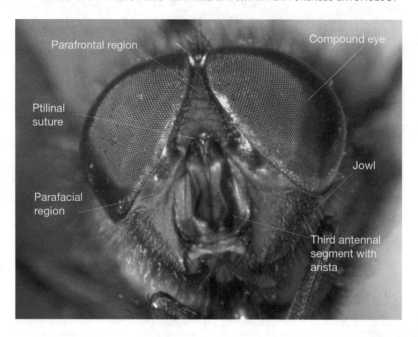

Figure 2.9   The face of a fly. The remains of the ptilinum is present as a ptilinal suture above the antennae

4. *The Muscophora Schizophora Calyptratae.* A division is made depending upon whether the halteres are exposed or covered. The opaque flaps concealing the halteres are termed the *squamae* or calypters; working from the body outwards, the proximal (near) flap is termed the lower thoracic squama and the distal (far) flap is called the upper thoracic squama. Flies which have covered halteres also have a complete line, or suture, across the prothorax and prominent thoracic protrusions, called *posterior callus*, sticking out from the middle of the thorax. Such fly species may also have a slit or cleft in the second antennal segment (working outwards from the head). According to Watson and Dallwitz (2003), however, this is considered an unreliable feature for identification purposes. These flies also have a ptilinum. On the top or dorsal side of the antenna there is a feathery protuberance, called the *arista*, which stands out from the antennal segment. Those flies with flaps covering the halteres and the features described above are termed *calyptrate*. Of particular forensic importance in this grouping are the families Calliphoridae, Sarcophagidae, Fannidae and Muscidae.

2.1.1   How to sex flies

The adult head has two large *compound eyes*. Those of the female blowfly are more widely spaced than those of the male, as you look at them head-on from the front. Another important feature is *occipital dilation* of the eyes. To see if this feature is

Figure 2.10   The occipital dilation present in *Calliphora uralensis* Villeneuve (A colour reproduction of this figure can be found in the colour section towards the centre of the book)

present, look at the fly head from the side view. It is possible to see if the eye is
expanded. This occipital dilatation is used to distinguish some species (Figure 2.10).

The region below the eye, as you look at the fly face side-on (lateral view), is
called the jowl. In *Calliphora vicina,* a species of forensic importance, this has a
mass of golden coloured hairs on it.

The families and the identification characteristics of some of the important
species are described in the next section. However, to ensure that you have correctly
identified a specimen, you should use keys, and have confirmed your identification
using a collection of named species and the diagnostic and differential description
in the handbook. Or you should ask a taxonomist to check your identification so
that you are absolutely certain of the name of the species. On this identification
hangs the determination of the post mortem interval.

2.2   Forensically important families of flies

2.2.1   Calliphoridae

The particular species of fly which are forensically important will differ from location to location. The first three species listed below are common initial colonizers
of corpses in Europe, including Britain, which are not buried or in some way 'modified'. For example, Schroeder *et al.* (2003) consider *Calliphora vicina, Calliphora
vomitoria* and *Lucilia sericata* to be the most common species found on corpses
in Germany.

Calliphora vicina (Robineau-Desvoidy)

This is a large blowfly, 9–11 mm in length. It is also recorded in the older literature as *Calliphora erythrocephala* (Meigen). The front thoracic spiracle is orange in colour (Smith, 1986). The head is black on top and the front half of the cheek (bucca) is reddish orange. The lower region of the face is black. There are black hairs on the jowls, irrespective of the jowl colour. The thorax is black and the top of the thorax (the dorsum) is covered with a dense greyish shine (*pubescence*). There are a pair of strong bristles in a row in the centre of the thorax. These are called the *acrostichal bristles* (Figure 2.11). Like other blowfly species, this species also has a fan of bristles, the *hypopleural bristles*, on a plate above the coxa of each hind (third) leg, near the *posterior spiracle*. Look for this spiracle and you will spot them. The abdomen is blue with a silvery chequerboard effect (*tessellation*) (Figure 2.12). The basicosta on the wing is yellowish in colour, although this can fade to a yellowish-brown colour.

Calliphora vomitoria (L.)

These are also large bluish-coloured blowflies. The species has a longer life cycle than the previous species and is more often found frequenting rural environments. The hairs on the jowls and the colour of the basicosta help identify *Calliphora vomitoria*. The basicosta is black in colour (Figure 2.13) (as opposed to the orange in *Calliphora vicina*) and the hairs on the base region of the jowls and around the

Acrostichal
bristles

Dorsocentral
bristles

Figure 2.11 The insect thorax showing the rows of acrostichal bristles down the centre of the thorax

Figure 2.12 The tessellation (chequer-board effect) on the abdomen of such flies as *Calliphora vicina* Meigen (A colour reproduction of this figure can be found in the colour section towards the centre of the book)

Figure 2.13 The black basicosta of *Calliphora vomitoria* (L.) (A colour reproduction of this figure can be found in the colour section towards the centre of the book)

side of the mouth are orange. The spiracle at the front (anterior) of the thorax is brownish in colour.

Where both *Calliphora vicina* and *Calliphora vomitoria* are found together as third instar larvae, they can be separated, according to Smith (1986), by the width of their posterior spiracles. He indicates that in *Calliphora vicina* the spiracles are 0.23–0.28 mm wide. The spiracles in this species are smaller than in *Calliphora vomitoria* and are separated by the same, or a bigger, distance than the width of a single spiracle. In *Calliphora vomitoria* the spiracles are larger, being in the region of 0.33–0.38 mm. Its spiracles are separated by less than the diameter of an individual spiracle.

Lucilia sericata (Meigen)

This is commonly called a greenbottle because all the flies in this genus are a metallic green colour. In North America, *Lucilia sericata* is called *Phaenicia sericata*. *Lucilia* species are distinguished from other blowflies by having a ridge just above the squama, the rear wing flap (hence the suprasquamal ridge), which has tufts of hair on it. *Lucilia sericata* has a yellow-coloured basicosta (Figure 2.14).

One of the differences between the larvae of *Calliphora* and *Lucilia sericata* is that the oral sclerite in the head skeleton (*cephalopharyngeal* skeleton) is transparent and so seems to be absent in larvae of *Lucilia sericata* (Figure 2.15a). The identity of *Lucilia sericata* larvae can also be confirmed by looking at the

Stem vein is
bare in Lucilia
species

Figure 2.14   The bare subcosta (stem vein) in *Lucilia* and *Calliphora* spp. distinguishes them from *Phormia* spp. where the subcosta upper surface is hairy (A colour reproduction of this figure can be found in the colour section towards the centre of the book)

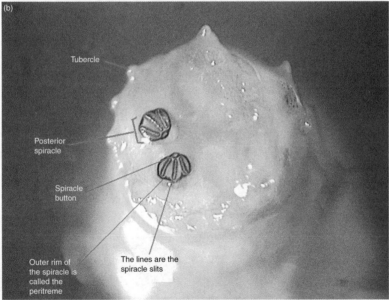

Figure 2.15 (a) An example of the head skeleton and (b) the tubercles and spiracles at the posterior region of the larva (A colour reproduction of this figure can be found in the colour section towards the centre of the book)

rim of the final posterior segment of the larva. The protrusions found along the outer rim of the segment are called *tubercles* (Figure 2.15b). They are named, from the top (12 noon position), the inner, median and outer (lower) tubercles. If the distance between the two inner tubercles is the same as the distance between the inner and the median tubercle, then this species can be identified as *Lucilia sericata*. This feature is characteristic of the third instar larvae. Erzinçlioğlu (1987) found that around the posterior spiracles in first and second instar larvae of *Calliphora* and *Lucilia* there was a circle of hairs. In *Calliphora* species these hairs would be visible under low power, being very well-developed in *Calliphora vomitoria*, but would not be visible under low power in *Lucilia* species.

Lucilia illustris (Meigen)

The *basicosta* is blackish or brown in colour in this species and the arista on the antenna has up to 10 hairs on its underside. There are no bristles on the sides of the abdominal segments in the males of this greenbottle species, according to Erzinçlioğlu (1996). The males can be distinguished from *Lucilia caesar* (Linnaeus) males, by the presence of curved surstyli (exterior structures of the genitalia) (Figure 2.16a). This fly was found to be of value as a post mortem indicator in a murder in Washington State (Lord *et al.*, 1986).

Lucilia caesar (L.)

These flies are similar to *Lucilia illustris* in that they share a dark-coloured basicosta. In males the sides of the second abdominal segment lacks bristles, when you look at the fly from the dorsal view (i.e. from the top). They can also be

(a)   (b)

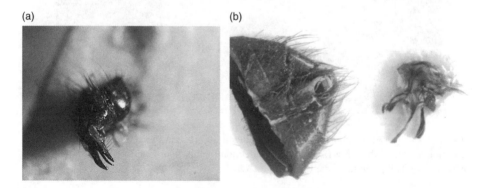

Figure 2.16 (a) *Lucilia illustris* Meigen surstyli. (b) *Lucilia caesar* (L.) surstyli (A colour reproduction of this figure can be found in the colour section towards the centre of the book)

distinguished by having surstyli with a straight projection (Figure 2.16b), i.e. fork (Erzinçlioğlu, 1996).

In the Soham, Cambridgeshire, murder in 2004, *Lucilia caesar* was found on the remains, although this was never used in court to identify the post mortem interval (Hall, personal communication).

Lucilia richardsi Collin

In this fly the basicosta is white or yellowish, and the spacing of the eyes in the male assists in distinguishing this species from adult *Lucilia sericata* (Figure 2.17). The distance between the eyes in males is not more than the width of the third antennal segment (Erzinçlioğlu, 1996). The abdominal sternites are hairy in both males and females (Greenberg and Kunich, 2002). Smith (1986) indicates that the tibia of the middle leg has two anterior bristles, which also distinguish this species from *Lucilia sericata,* which has only one bristle (the leg articulates in the posteroventral plane (i.e. backwards and downwards).

Protophormia terraenovae Robineau-Desvoidy

This species is 8–12 mm long. The fly has a greenish-blue abdomen, black-coloured legs and a dark calypter with hairs which are dark (Figure 2.18). According to

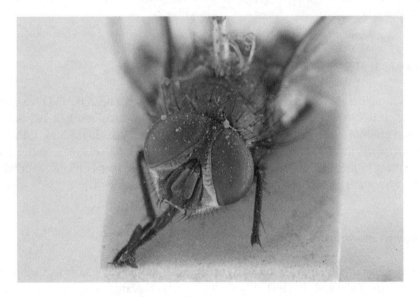

Figure 2.17 *Lucilia richardsi* Collin (A colour reproduction of this figure can be found in the colour section towards the centre of the book)

Figure 2.18 *Protophormia terraenovae* (Robineau-Desvoidy) (A colour reproduction of this figure can be found in the colour section towards the centre of the book)

Smith (1986), this species is widely distributed in Britain and its puparia may be recovered from the body, rather than at some distance from it, according to Busvine (1980). Tantawi and Greenberg (1993) provide information about the lengths of the life stages of *Protophormia terraenovae* at 12.5°C, 23°C, 29°C and 35°C.

Phormia regina (Meigen)

This is a smaller fly than those previously described and is a nearctic and palaearctic species. It is 7–9 mm long and has a green or greeny-olive coloured body. Its head is large proportional to the body and is black in colour. A distinguishing feature in this species is the anterior spiracle on the thorax, which has obvious orange hair. In contrast to that in *Protophormia terraenovae*, the calypter is white, with white hairs. *Phormia regina* is commonly known as the black blowfly.

Cynomya mortuorum (Linnaeus)

This is a metallic blue-green blowfly which is about the same size as *Calliphora* species. Its face and jowls are yellow to bright orange (Figure 2.19). It is infrequently found in the south of England and MacLeod and Donnelly (1956)

Figure 2.19 *Cynomya mortuorum* (L.), illustrating its yellow parafrontal region (A colour reproduction of this figure can be found in the colour section towards the centre of the book)

note that it favours cool uplands. The larvae of *Cynomya mortuorum* are associated with unburied corpses.

Chrysomya species – Chrysomya rufifacies (Macquart)

These are large blue or green flies. *Chrysomya rufifacies* is most commonly found in the orient, Australasia and the neotropics. It is metallic bluish or green in colour. The adults are 6–12 mm long, with at least the front part of the cheeks on the head being yellow or orange in colour (Smith, 1986). *Chrysomya rufifacies* is one of the initial colonizers of corpses in Hawaii (Goff, 2000). Its larvae have spines on the sides of their tubercles.

Chrysomya rufifacies is often accompanied by *Chrysomya megacephala* (Fabricius), which is of a similar size. In contrast to *Chrysomya rufifacies*, the anterior spiracle of *Chrysomya megacephala* is orange to black-brown in colour, rather than being white to pale yellow. The front part of the cheeks (bucca) in this species is yellowish or orange. *Chrysomya megacephala* has also been identified from corpses along with *Cochliomyia macellaria* (Weidemann), a native American species in Brazil (Oliveira-Costa and de Mello-Patiu, 2004).

Chrysomya albiceps Wiedemann is a third species which is found at crime scenes. It has a yellowish or white thoracic spiracle, its abdomen has dark bands across it and its legs are dark. Larvae of *Chrysomya rufifacies* and *Chrysomya albiceps* are hard to distinguish visually. However, Wells and Sperling (1999) demonstrated that the two species could be distinguished by using mitochondrial DNA.

2.2.2 Sarcophagidae

The common name for a member of this family of flies is the fleshfly. They are large and greyish in colour and have a chequerboard (tessellated) abdomen which is silvery grey and a thorax with three stripes down it (Figure 2.20). [In volume 12 (Chandler, 1998) of the *Checklists of Insects of the British Isles (New series), Part 1: Diptera*, it is noted that there has been a division of the old genus *Sarcophaga* into a number of subgenera, resulting in a number of name changes from the early literature. It may be helpful to consider this in your reading of the earlier scientific papers].

Colyer and Hammond (1951) considered this a difficult group from which to identify species with any degree of certainty, unless adult specimens are captured whilst mating ('in cop'). Using the identity of the male, it is easier to confirm the identity of the female species. Help should be sought from taxonomists if these are the only family recovered from the body.

The sarcophagid larvae are characterized by having a barrel-like shape with their posterior spiracles sunk into a hollow. The edge of the posterior segment has a large number of tubercles. This makes this family easy to distinguish as a larval stage. Some success has also been made in identification to species of larvae of the Sarcophagidae, using molecular methods (Zehner *et al.*, 2004).

Figure 2.20 A fly showing the characteristic features of the Sarcophagidae (A colour reproduction of this figure can be found in the colour section towards the centre of the book)

2.2.3 Sepsidae

Sepsid flies are small and a shiny black colour. They have a head which looks spherical with bulging (convex) eyes and an abdominal constriction (an apparent waist!). The legs of male sepsids have spines and are elongated, which makes the legs look deformed. The costa on the wing is unbroken. This family is characterized by its habit of wing-waving. There may be swarms of these flies at the crime scene, depending on its location. Pont and Meier (2002) have revised the European Sepsidae and provide further details of their distribution as well as details of characteristics for identification.

2.2.4 Piophilidae

The Piophilidae are small, shiny, black flies (Figure 2.21), 2.5–4.5 mm in length. The costal vein of the wing appears broken at one point in this family. One of the most researched members of the Piophilidae is *Piophila casei* (Linnaeus), which is a food pest on such products as cheese.

Piophila casei (L.)

This fly is usually found on the corpse at the end of active decay and the start of the dry stages (Byrd and Castner, 2001). It is a small black fly, 2.5–4.0 mm

Figure 2.21 A pair of piophilids taken 'in cop'

Figure 2.22   Head of *Piophila casei* (L.) to show the cheek region (A colour reproduction of this figure can be found in the colour section towards the centre of the book)

in length, and is commonly called the cheese skipper. It has prominent cheeks which are more than half the eye height (Figure 2.22). These flies have a yellow colour on their legs, antennae and on the jowls of their faces. Their ocellar bristles are found opposite the simple eye (front ocellus), which are small and widely spaced.

Piophilid larvae are similar to sepsid larvae, although the posterior larval region is narrower in the Piophilidae. The behaviour of *Piophila casei* larvae makes them easy to identify, as it is particularly characteristic. If disturbed, the larvae bend round to grasp two small papillae on the posterior segment, using their mouth hooks. They suddenly release the papillae and the larvae unexpectedly 'jump' up to 15 cm into the air.

Other species of Piophilidae may also be present on corpses. For example, *Stearibia* (= *Piophila*) *nigriceps* (Meigen) was recorded by Oldroyd (1964) as feeding on a human corpse. Therefore, do not make an assumption about the name of the species of piophilid which you have recovered from the body.

2.2.5   Phoridae

These are small, often minute, flies. They have a humped back and can be greyish-brown or bluish in colour. The forehead (*frons*) is usually wide, and has bristles which are very robust and upward curving. On the antenna the third segment is large, although in this family the arista can be found either dorsally or apically (frontward). Phorid wings are characteristic; with veins 1–3 appearing very pronounced and crushed together. The wing costa also has a spine at its proximal end, nearest the body of the fly.

On a corpse, Phoridae can be identified by the fact that they are active flies, capable of running and jumping, and this gives them their common name of scuttle flies. Dewaele and LeClerq (2002) define their flight period as April to November.

2.2.6 Muscidae

Muscid flies are frequently greyish in colour and are characterized by having lines running down the length of their thorax and no hypopleural bristles. Wing veins 6 and 7 are short and do not move towards each other (as they do in *Fannia* spp.). Their squamae are of roughly the same size, or the lower squamae may be bigger (Unwin, 1984).

Box 2.2   Hint The DNA molecule, mitochondria and insect mtDNA

DNA

The deoxyribonucleic acid (DNA) molecule is made of genetic material, stored as base pair sequences – adenine (A), guanine (G), thymine (T) and cytosine (C). As a rule of thumb, three DNA base pairs make a code for an amino acid. The DNA molecule is a double helix of two intertwined DNA strands (the nuclear DNA is diploid).

Mitochondria are the organelles in which ATP is predominantly manufactured. They are 0.5–1.0 μm in diameter and 5–10 μm in length, and have an outer membrane surrounding an inner membrane with lots of infolding (cristae). This allows compartmentalization of the mitochondrial biochemical processes. The chemicals responsible for respiration electron transfer are bound to the inner membrane of the mitochondria.

Knowledge of insect DNA is based on research on the fruit fly. This has shown that the insect mtDNA has a large number of adenine and thymine bases (it is A + T-rich) found in a non-coding region (the control region) and genes coding for cytochrome oxidase subunits I and II, amongst others. The control region is also likely to be where the initiators for replication are located.

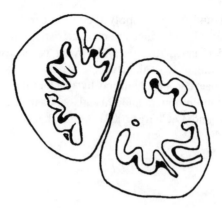

Figure 2.23   The sigmoid spiracle slit shapes characteristic of Muscidae

Muscid flies, such as *Musca domestica,* a common member of the Muscidae, will visit a body soon after death, attracted by any exudates rather than the corpse itself. *Musca domestica* is a greyish fly, about 6–7 mm in length. It is characterized by four narrow black stripes along its thorax and a greyish or yellowish abdomen. A sharp-angled wing vein is found at vein 4 (Smith, 1986).

Musca autumnalis De Geer is also recorded visiting corpses (Smith, 1986). This is commonly called the 'face fly' and the male is easily identified by its bright yellow abdomen with a black stripe up the middle. According to Smith (1986), female *Musca autumnalis* are very similar to female *Musca domestica,* but can be distinguished by a smaller frontal stripe. In *Musca autumnalis* this stripe is less than twice the width of the eye, whilst in *Musca domestica* it is three to four times as wide. *Musca autumnalis* is rare in the north of England up to the borders of Scotland and is not found in Ireland.

The larvae of the Muscidae are recognizable by the wiggly S-shaped slits on the posterior spiracles (Figure 2.23).

2.2.7 Fannidae

These flies are commonly called latrine flies. They are distinguished from the muscid flies by the much greater curve on the axillary vein (vein 7; this is the vein which is nearest to the upper calypter) and the species also lacks the sharp angle on vein 4, which reaches to the wing margin. *Fannia canicularis* (Linnaeus), the lesser house fly, is common in houses as it is attracted to light. Smith (1986) comments that this is the more common house fly until July. Benecke and Lessig (2001) recorded *Fannia canicularis* in a case of child neglect in central Germany. They suggested that urine and faeces attracted the adult flies, resulting in larval infestation of the child's genitalia.

2.2.8 Sphaeroceridae

Sphaeroceridae are known as small dung flies. They are dull-coloured flies, 1.5–5 mms in length. Their antennae are three-segmented with an arista. *Vibrissae* (bristles) are present on the sides of the mouth. The wing costa has two breaks and vein 6 is present, but does not extend to meet the wing margin.

Sphaeroceridae filter-feed on bacteria. Only a few species in this family are recorded from dead bodies. They have been noted from the fresh, bloat and advanced stages of decay (the fifth wave of insects) at 4–8 months after death. Grassberger and Frank (2004) recorded them on dressed pig corpses placed in an urban garden in Vienna between May and November 2001. Ammonia is an attractant for the dung-breeding flies. Hence, voiding of urine early in decomposition can attract members of this family as much as ammonia release during later decomposition.

2.3 DNA identification of forensically important fly species

Identification of the species of some larval instars can be difficult using morphological features. Molecular biology has a role to play in providing an alternative means of insect identification using DNA. As in human somatic cells, insect cells contain genetic material in two places. DNA is found in the nucleus and is called nuclear DNA. It is also found in the cytoplasm as small circular structures in organelles called mitochondria. Both of these types of DNA are extracted in order to identify the life stages of insects, such as flies, collected from a body. Where the entire DNA content of the cell is used, it is described as 'genomic DNA'. If only the DNA from the mitochondria is utilized, then the DNA is referred to as 'mitochondrial DNA' (or, in shorthand form, as 'mtDNA'). Mitochondrial DNA is easier to extract than nuclear DNA because there is more of it available.

Insect species identification using either mtDNA or nuclear DNA is based on the sequences of nucleotides on the chromosomes. These sequences are called loci and are made up of an arrangement of sections of strings of base pairs of the nucleotides adenine, thymine, cytosine and guanine (A, T, C and G) which form the DNA molecule. The sections of base pairs which are used may be very short.

Where they are made of fewer than 1000 base pairs, the length of DNA has to be artificially increased or '*amplified*' before it can be interpreted. The method used to increase the DNA sections is a process called *polymerase chain reaction* (PCR).

To replicate the required regions of sample DNA, previously generated sections are joined to the sample DNA at known sites, to enable it to be copied. These sections are called *primers*. Specific primers have been generated to identify calliphorid DNA (Table 2.1). This means you are 'amplifying' a PCR product

Table 2.1   Examples of primers for cytochrome oxidase investigation of calliphorid identity

Primer title	Composition	Reference source
CO-I 2f	5'-CAG CTA CTT TAT GAG CTT TAG G-3'	Vincent, Vian and Carlotti, 2000
CO-I 3r	5'-CAT TTC AAG C/TTG TGT AAG CATC-3'	Vincent, Vian and Carlotti, 2000
TY-J-1460	TAC AAT TTA TCG CCT AAA CTT CAG CC	Wells and Sperling, 2001
C1-N-1687	CAA TTT CCA AAT CCT CCA ATT AT	Wells and Sperling, 2001
C1-J-2319	TAG CTA TTG GAC/TTA TTA GG	Wells and Sperling, 2001
C1-N-2514	AAC TCC AGT TAA TCC TCC TAC	Wells and Sperling, 2001
C1-J-2495	CAG CTA CTT TAT GAG CTT TAGG	Also used by Harvey, Dadour and Gaudieri, 2003
C1-N-2800	CAT TTC AAGT/CTG TGT AAG CATC	Also used by Harvey, Dadour and Gaudieri, 2003

from a known site, so you can identify the nucleotides and their position on the DNA molecule.

Some species identifications are based upon short sections (loci) on the DNA molecule, which can be fewer than 350 significant base pairs. This means that, although the DNA chain degrades over time, insects which are stored, or which have dried out, can still be reliably identified using suitable primers.

Both killing and preservation methods used at the crime scene, and subsequently, can influence the quality of the DNA which is recovered. Dillon *et al.* (1996) pointed out that the use of ethyl acetate to despatch specimens later intended for DNA extraction could reduce the amount of DNA extracted, although other workers disagree (Fukatsu, 1999; Dean and Ballard, 2001). Using some preservatives can result in fragmentation of the DNA molecule. Preservation of specimens in 99 % alcohol provided fragments of up to 1400 base pairs, according to Sperling *et al.* (1994). In contrast, flies stored dry, or preserved in 75 % ethanol, provided DNA fragments reduced to up to 350 base pairs.

Any insect specimens chosen for DNA extraction should be taken from live cultures and killed by freezing. Freezing adult flies immediately at −70°C ensures that the DNA has not degraded as rapidly as it might be if other methods of preservation were used. However, work by Lonsdale *et al.* (2004) indicates that the duration of frozen storage can also affect the degree of degradation of the DNA molecule when the period of storage time is greater than one year.

Once killed, the specimens need to be cleaned before the DNA is extracted, to remove contamination by foreign DNA. Washing initially in bleach or acetone has been found to be satisfactory and does not affect the level of DNA preservation (Fukatsu, 1999; Linville and Wells, 2002). Linville and Wells (2002) found that washing maggots in a solution of 20 % bleach reduced contamination without material in the crops of the larvae being affected. A second precaution against contamination is to analyse the DNA from the head or thorax of the individual adult fly or the mid-section of a larva. This allows, as is necessary in all forensic work, the retention of voucher specimens in the form of the remaining body parts. Where possible in forensic work, the post-feeding stage should be used, or the larvae should be starved so that their gut is empty of food. This, too, prevents contamination by DNA, other than that of the particular individual species being investigated.

The choice of extraction chemicals varies between laboratories. Junqueira *et al.* (2002) concluded that DNAzol® was the most effective chemical for extracting DNA, compared to extraction using Chelex® or the phenol/chloroform method, particularly if there was a fear that the DNA could be damaged.

Once extracted, the fly DNA can be investigated further in one of three main ways: *random amplified polymorphic DNA* (RAPD); *mitochondrial* DNA (mtDNA); and *restriction fragment length polymorphism* (RFLP). These will each be discussed under separate headings.

2.3.1 Random amplified polymorphic DNA (RAPD)

This method uses non-specific primers and the PCR products come from many areas of the specimen sample of DNA. Benecke (1998), for example, suggests using Primer 5 and REP 1R for forensic case work. Based on a 5′–3′ sequence, the RAPD primers are:

REP 1R*   XIIIACGTCGICATCAGGC
Primer 5   XAACGCGCAAC

(*from a description provided by Pharmacia Biotech and referenced in Benecke, 1998).

The results are read as peaks to give a peak profile, or 'signature' (Figure 2.24). Only the strongest peaks, those of maximum intensity of 50 % and over, are used to

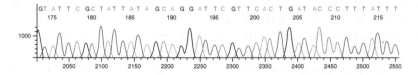

Figure 2.24 An example of an electropherogram (A colour reproduction of this figure can be found in the colour section towards the centre of the book)

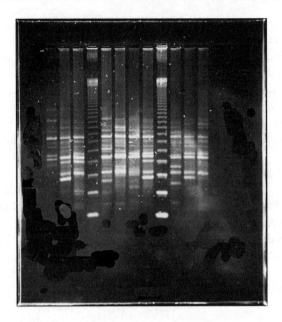

Figure 2.25   Electrophoresis gel from RAPD analysis of fly DNA

generate the profiles for the individual fly species. If the numbers of strong peaks are few, Benecke suggested repeating the work using more primers. If the peak tops split, he suggested switching from the electrophoretic display (Figure 2.24) on the sequencer to a silver stain-like banding pattern view. Benecke also pointed out that, to avoid getting false positives, samples from the same DNA source should never be loaded next to each other on the electrophoresis gel (Figure 2.25).

The advantages of using RAPD analysis are many. For example, because commercially generated primer beads are available, the process is quick; the room for experimental error and sample contamination during the analysis is reduced; the chemicals for RAPD analysis have a long shelf-life; and a large amount of information is recovered.

Stevens and Wall (1995) used RAPD analysis to investigate variation between populations of *Lucilia sericata* from farms near Weybridge in Surrey and those from farms in the Bristol area. As a control, they included specimens of *Lucilia sericata* from a laboratory culture at the University of Bristol. The results showed that closely related individuals of the species *Lucilia sericata* could be distinguished on the basis of their RAPD profiles.

There are some disadvantages of using RAPD analysis of insect DNA, however, since the signatures gained for different species will not be standardized; no national or international databases exist from which to compare RAPD profiles of the insect species, and no statistical data are available in order to exclude chance when interpreting the results. For these reasons, RAPD results are only used for presentation to a court where the data for the specimens show indisputable differences.

2.3.2 Mitochondrial DNA analysis

Mitochondria are tiny organelles in the cytoplasm, where a stage of respiration called oxidative phosphorylation takes place. As a result, ATP is generated, using enzyme complexes called cytochromes, including cytochrome *c* oxidase, an enzyme complex made up of three subunits. This enzyme complex (complex IV) is found in the inner membrane of each mitochondrion and is the third and final enzyme for the electron transfer chain involved in mitochondrial oxidative phosphorylation.

The insect mitochondrial DNA is a small circular genome containing around 16 000 base pairs of double-strand DNA, which comes predominantly from maternal sources (Lessinger *et al.*, 2000). The molecule comprises approximately 37 genes (22 for transfer RNA, 2 for ribosomal RNA and 13 for peptides). These genes include those for two subunits of cytochrome c oxidase, subunits I and II (COI and COII). COI was originally chosen by molecular biologists to investigate genetic profiles, because it is the biggest of the three mitochondrially encoded cytochrome oxidase subunits and the protein sequence combines both variable and highly conserved regions (Morlais and Severson, 2002, quoting Clary and Wolstenholme, 1985; Beard *et al.*, 1993; Saraste, 1990; Gennis, 1992).

Box 2.3 Hint Cytochrome oxidase

Cytochrome oxidase

In respiration, cytochrome oxidase is a protein that acts as an oxidizing enzyme involved in the transfer of electrons from cytochrome to oxygen molecules. It has two catalytic subunits, COI and COII. Cytochrome oxidase is made up of:

- Cytochrome *a*

- Cytochrome *b*

- Two copper atoms

- Thirteen individual subunits of protein (three of which are encoded for by mtDNA)

In insects, the non-coding region of the mtDNA structure is called the control region or the A-T region. This is because it is rich in the nucleotides adenine and thymine and also controls replication of mitochondrial DNA and the transcription of RNA (Avise *et al.*, 1987). To describe the sequence of base pairs, so that an individual 'signature' or *haplotype* can be specified for a particular species, a nucleotide position numbering system is used. This follows that described for the fruit fly *Drosophila yakuba* (Burla) (GenBank Accession No. NC–001322).

Mitochondrial DNA (mtDNA) is useful for insect species identification as it is, for the most part, resistant to degradation and its use can enable forensic scientists to provide identification of fly species within a day. The mtDNA mutation rate, according to Avise (1991; cited by Sperling *et al.*, 1994), is such as to be able to distinguish between closely related species of insects. For example, it was effectively used by Avise *et al.* (1987) to distinguish between *Phormia regina*, *Phaenicia (Lucilia) sericata* and *Lucilia illustris*.

Details of primer sequences, specifically relating to sequences of mtDNA in Calliphoridae, with the reference numbers L14945–7, can be accessed from GenBank (Malgorn and Coquez, 1999). This information can then be used to request prepared primers from pharmaceutical companies. The enzymes used are robust at a range of temperatures, can be used with various buffers and are moderately inexpensive, at approximately £2.50 per sample. Ideally, primer enzymes that give a breadth of potential amplification should be used.

Once extracted, the sequences of the protein-coding regions of the mtDNA, e.g. those coding for cytochrome oxidase subunits 1 and 2 (COI and COII), are compared with known species 'signatures' in the Genbank database using computer software. GenBank is a database of genetic profiles that are ratified and publicly available. Software such as Blast Search, which is accessed via the web (www.ncbi.nlm.gov/), is used to search GenBank.

Small intra-specific differences in the sequences of individual larvae from the same species do occur, although the overall 'signature' is constant. For example, Wells *et al.* (2001) recorded a substitution of adenine at position 2058 (with reference to the *Drosophila yakuba* numbering system). Because the resulting amino acid sequence is unchanged, this was called a 'silent substitution' and considered typical of cytochrome oxidase subunit *b* (COI) haplotype variation (where there is more than one haplotype, because of mutation in single organisms of a particular species, the condition is termed *heteroplasmy*).

A case study of the use of mtDNA analysis

Wells *et al.* (2001) showed that DNA from the corpse could be recovered from maggots using mtDNA. A live donor, who had had his liver removed for a transplant, was used in this research. The fly larvae were fed on the donor's excised liver and their gut contents were then analysed. A control was provided by analysing a blood sample from the human transplant patient. DNA from both sources had the same characteristics

Wells *et al.* utilized this same technique of investigating the larvae to try to confirm the previous presence of a corpse at a crime scene. In 1989 a collection of maggots, subsequently identified as *Chrysomya albiceps*, was found on the floor of a farmhouse cellar in southern Italy. The police, in response to a 'tip-off', had made a raid on the farmhouse expecting to find a body in the cellar, but there was no body. This was because in the interim the perpetrators had discovered that they had been betrayed and moved the corpse. Entomologically-aware police officers collected the maggots from the floor of the cellar to try to link this location with the earlier presence of the victim. The DNA content of the larval crops of the blowflies was analysed, together with material originating from the missing body. The scientists confirmed that both samples of DNA were from the same sources.

2.3.3 Analysis of insect DNA using restriction fragment length polymorphism (RFLP)

This is a faster and less expensive method than sequencing DNA using mtDNA. However, in this instance, restriction sites, which are fixed for a particular species, have to be determined before using the technique. This is a problem, since mutations can arise which affect the DNA signature and make its interpretation difficult. In methodological terms it is helpful to use a range of six 'pre-prepared' primers to establish similarity coefficients and, wherever possible, to use post-feeding larvae to avoid contamination of the DNA sample by the food the maggots had eaten previously, to overcome this problem.

RFLP–PCR methods have been successfully used in forensic cases to assist in species identification, to allow post mortem interval determination. Schroeder *et al.* (2003), to undertake RFLP analysis, used a modification of the method described by Sperling *et al.* (1994) for analysing mitochondrial DNA. They separated species of *Calliphora vicina*, *Calliphora vomitoria* and *Lucilia sericata* found on human corpses and successfully sequenced a 349 bp section of the mtDNA from parts of the cytochrome oxidase gene for subunit I (COI), the cytochrome oxidase subunit II gene (COII) and the tRNA leucine gene. From these specific regions they clearly distinguished the species which are the three most common corpse-infesting flies in the Hamburg area.

Similar work was undertaken using PCR–RFLP to identify three species of blowfly characteristic of Taiwan. Chen and Shih (2003) confirmed that the partial gene of subunit cytochrome oxidase I ITSI region of ribosomal DNA (rDNA) could be used to successfully distinguish between *Chrysomya megacephala*, *Chrysomya pinguis* (Walker) and *Chrysomya rufifacies*.

DNA is not the only molecular means that has been used to characterize flies. Other chemicals, including allozymes, have been used for this purpose.

For example, by extracting 22 cuticular hydrocarbons and using GC–MS, Byrne *et al.* (1995) identified a distinction between geographic populations of *Phormia regina*. These workers discriminated, on the basis of amount of hydrocarbons, between both the sexes and the three different geographic locations sampled.

2.3.4 The use of allozymes to identify species of fly

Allozymes are enzymes which, because of a genetic mutation, can show variations in individual species. Differences in enzymes can be investigated by using electrophoretic techniques such as isoelectric focusing and isolating the proteins as bands on polyacrylamide gels.

Allozymes have been used in southern Australia to identify a range of blowflies commonly found associated with bodies (Wallman and Adams, 2000). Four southern Australian species of fly were investigated. These were *Calliphora dubia* (Macquart), *Calliphora stygia* (Fabricius), *Calliphora hilli hilli* (Paton) and *Calliphora vicina*. Using 42 allozymes, Wallman and Adams were able to show a clear distinction between the four species, both at third larval instar stage and as adults.

The main advantage of this method is its speed. Wallman and Adams were able to generate results in 3 hours, which is more rapid than is possible for DNA analysis. This method of analysis is also cost-effective and has a high level of reliability. Loxdale and Lushai (1998) suggest that the cost of each sample is around £0.15. They also point out that the method can be carried out in the field using equipment powered by portable battery packs.

A disadvantage of the use of allozymes, which is often cited, is the production of weak banding patterns. Loxdale and Lushai suggest that this can be overcome by applying the sample accurately to the gel plate, investigating the buffers so that the most effective is used and keeping the reaction cool whilst electrophoresis is carried out. Further studies need to be undertaken in order to develop this methodology, so that diagnostic markers for forensically important fly species can be reliably defined and the limits of variability can be confirmed, before this technique can have a natural place in the court room.

2.4 Further reading

Byrd J. H. and Castner J. L. (eds). 2001. *Forensic Entomology: The Utility of Arthropods in Legal Investigations.* CRC Press: Boca Raton, FL.

Campobasso C. P., Linville J. G., Wells J. D. and Introna F. 2005. Forensic genetic analysis of insect gut contents. *American Journal of Forensic Medicine and Pathology* **26** (2): 161–165.

Cruickshank R. H. 2002. Molecular markers for the phylogenetics of mites and ticks. *Systematic and Applied Acarology* **7**: 3–14.

Erzinçlioğlu Y. Z. 1985. Immature stages of British *Calliphora* and *Cynomya*, with re-evaluation of the taxonomic characters of larval Calliphoridae (Diptera). *Journal of Natural History* **19**: 69–96.

Erzinçlioğlu Y. Z. 1996. *Blowflies*. Naturalists' Handbooks No. 23. Richmond Publishing: Slough.

Gaudry E. 2002. Eight squadrons for one target: the fauna of cadaver described by P. Mégnin. Proceedings of the first European Forensic Entomology Seminar, 77 eafe.org/OISIN_2002; pp 23–28.

Harvey M. L. 2005. An alternative for the extraction and storage of DNA from insects in forensic entomology. *Journal of Forensic Sciences* **50**(3): 1–3.

Harvey M. L., Dadour I. R. and Gaudierie S. 2003. Mitochondrial DNA cytochrome oxidase I gene: potential for distinction between immature stages of forensically important fly species (Diptera) in western Australia. *Forensic Science International* **131**: 34–139.

Mullen G. and Durden L. (eds). 2002. *Medical and Veterinary Entomology*. Academic Press: Amsterdam.

Pont A. C. 1979. Sepsidae: Diptera, Cyclorrhapha, Acalypterata. *Handbooks for the Identification of British Insects* **X**(5c): 1–23.

Rognes K. 1991. Blowflies (Diptera: Calliphoridae) of Fennoscandia and Denmark. *Fauna Entomologica Scandinavica* **24**.

Sukontason K., Sukontason K. L., Piangjai S., Choochote W. *et al.* 2004. Fine structure of eggs of blowflies *Aldrichina grahami* and *Chrysomya pacifica* (Diptera: Calliphoridae). *Biological Research* **37**: 483–487.

Sukontason K., Sukontason K. L., Piangjai S., Boonchu N. *et al.* 2004. Identification of important fly eggs using potassium permanganate staining technique. *Micron* **35**(5): 391–395.

Unwin D. M. 1984. *A Key to the Families of British Diptera*. AIDGAP, Field Studies Council Publication No. S9 (now No. 143). Headly Brothers Ltd: London [reprinted from *Field Studies* **5**: 513–553 (1981)].

Walman J. F. 2001. A key to the adults of species of blowflies in southern Australia known or suspected to breed in carrion. *Medical and Veterinary Entomology* **15**: 433–437.

Useful websites on fly morphology and identification

Look within the web page www.ento.csiro.au/biology/fly/fly.html for the link to flyGlossary.html to gain access to an interactive approach to identifying the key features of fly anatomy.

Other web addresses of interest include:

Key to Fly Anatomy
www.nku.edu/~biosci/CoursesNDegree/ForensicFlyKey/flyanatomy.htm

Identification key to Calyptrate Diptera families: www.nku.edu/~biosci/CoursesN Degree/ForensicFlyKey/families.htm

Key for identification of Calliphoridae to species: www.nku.edu/~biosci/CoursesN Degree/ForensicFlyKey/species.htm

3

Identifying beetles that are important in forensic entomology

3.1 What do beetles look like?

Beetles belong to the order Coleoptera and all share features in common. For example, they have biting mouthparts or *mandibles*, their antennae characteristically have 11 segments (although in some species there may be fewer than this) and the first section of the thorax (the prothorax) is usually distinctive in shape and size and can be used as an means of identifying the beetle. The beetle exoskeleton is formed from hardened plates. The plates on the top surface are called *tergites*, the plates on the under-surface (ventral) are called *sternites*. The segment plates at the side (lateral) of the body are called *pleurites* (the *pleuron* is the name for this region of the exoskeleton).

Beetle adults are composed of a head, a thorax in three parts all fused together (although the second and third parts are less visible dorsally) and an abdomen. They have two pairs of wings; the two forewings are hardened and form a protective covering over the second, membranous pair of wings. These chitinous, and on occasion 'leathery', protective cases are called the *elytra* (singular *elytron*).

The prothorax is well developed and, together with the head, can be interpreted as a distinct anterior section of the body. The dorsal surface of the thorax is divided into the *pro-, meso-* and *metanotum* (each plate, or tergite, is called a *notum*; plural *nota*). The *pronotum* (the surface of the first thoracic segment in front of the elytra) is the biggest of the thoracic segments. It is made up of only one plate (Figure 3.1). The ventral surface is correspondingly divided into three; the *pro-, meso-* and *metasternum*).

The middle region of the thorax (the mesothorax) supports a pair of hardened wing cases which meet along the centre of the dorsal surface of the body. Part of the mesonotum is located between the base of the elytra, behind the pronotum; this small plate is called the *scutellum*.

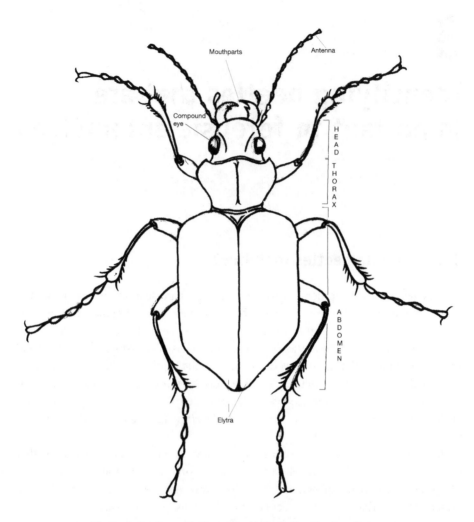

Figure 3.1   Generalized morphology of a beetle, dorsal view

The membranous wings are attached to the body on the last section of the thorax (the metathorax, with which the mesothorax is fused). This pair of wings is folded under the elytra when the beetle is not in flight.

Beetles' legs are positioned on the sternum. They are generally designed for running or walking, but in some beetles, as in the Scarabaeidae, the front legs are also modified for digging (Figures 3.2 and 3.3). The upper plates of the abdominal segments are *sclerotized* (made of hardened cuticle due to the formation of a protein called sclerotin). The lower abdominal plates (the sterna) are soft.

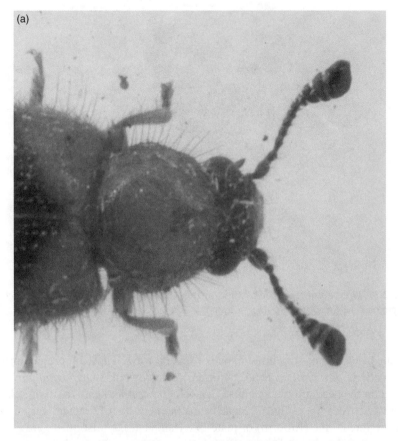

Figure 3.2 Structure of a beetle thorax; (a) dorsal view and (b) ventral view

Figure 3.3 Insect front legs modified for digging (*Geotrupes* sp.) (A colour reproduction of this figure can be found in the colour section towards the centre of the book)

Beetle heads can be structured in one of several ways. They can project forwards horizontally (a *prognathous* head), or orientate downwards (a *hypognathous* head). Located on the head are the antennae, which carry tactile, heat-sensitive, olfactory and humidity receptors. The antennae of members of the Coleoptera vary in form. Some are thread-like (filiform) or plate-like (lamellate), whilst others are elbowed (geniculate) or have club-like ends (clavate).

Beetles exhibit complete metamorphosis during their life cycles and pass through an egg stage, larval stages and a pupal stage and emerge as an adult, or imago, each of which is morphologically different. Beetle eggs are frequently difficult to locate on or around the body, as unlike fly eggs they do not often appear in batches on a body.

Beetle larvae have more distinctive morphological features than do the larvae of flies. For example, they have a sclerotized head capsule, and mouthparts which include mandibles (are mandibulate). Larvae may or may not have legs on the thoracic region of their body. *Prolegs* (limbs on the abdominal region) are rarely present in beetle larvae and this distinguishes them from the larvae of other orders.

For example, ground beetle larvae (carabids) have an elongated flattened shape with well-defined legs that end in two claws. These are called *campodeiform* larvae. Scarab beetle larvae resemble a C–shape and these beetles tend to have a brown sclerotized head and a whitish body. On the other hand, larvae of the dermestid family are particularly hairy on both the sides and the upper body surfaces and are recognized because of this coat of hairs. Examples of the shapes of forensically significant larvae are shown in Figure 3.4.

Box 3.1 Hint Structure of Cuticle

Cuticle

Insect *cuticle*, which is made of *chitin* and proteins, can be rigid or flexible. Cuticle provides protection from physical damage and water loss and a rigid structure for muscle attachment, and limits growth to those times when the cuticle is newly developing. The mechanical properties of cuticle depend on the quantity of protein present, the sequence of proteins and the degree of tanning (sclerotization).

Cuticle has three parts: epicuticle, procuticle and epidermis. The epidermis and cuticle together are called the insect *integument*. The *epicuticle* is the outermost layer. It is 0.1–3.0 μm thick and is also made of three layers. The outermost layer is a cement layer which prevents distortion of the next layer, a lipid–protein layer. Below this second layer is a glycoprotein superficial layer. The epicuticle does not contain chitin. It is not capable of providing support or extending, but does provide waterproofing and protection against mechanical damage.

Below this is the *procuticle*, which is 0.5–10 μm in depth and comprises a thicker endocuticle, which is light in colour, overlaid by a thinner, darker exocuticle. Procuticle is made up of a protein matrix in which layers of parallel microfibrils of chitin, an amino-sugar polysaccharide, are embedded to make a sheet. In the exocuticle the sheets of microfibrils are in the same plane, but each sheet may be orientated at a slight angle to the previous sheet. An alternate stacked or helicoid arrangement of microfibril sheets in the endocuticle results in it being a thicker layer than the exocuticle. The darkening of the thinner exocuticle is due to tanning (sclerotization).

The basal layer beneath the cuticle is the *epidermis*. This single layer of cells is supported on a basement membrane which separates the exoskeleton from the main body cavity. Epidermal cells regenerate by cell duplication, or mitosis. This layer secretes the cuticle-forming chemical which is needed for moulting to take place.

Types of cuticle

There are two types of cuticle, soft and hard:

1. *Soft cuticle* is flexible and the cuticle is thin and has little or no exocuticle. Larvae predominantly have soft cuticle and a hydrostatic skeleton. Soft cuticle is also important where movement is required and, for example, allows gravid females to extend their abdominal plates to lay eggs.

2. *Hard cuticle* is hardened and armour-like because of the level of tanning, the positions of the microfibril sheets and hydrogen bonding between adjacent chitin molecular chains. Hardened chitin is found surrounding the spiracles of fly larvae and is present on the head and as the mandibles of beetle larvae. It provides the strength and rigidity of the body and elytra in adult beetles.

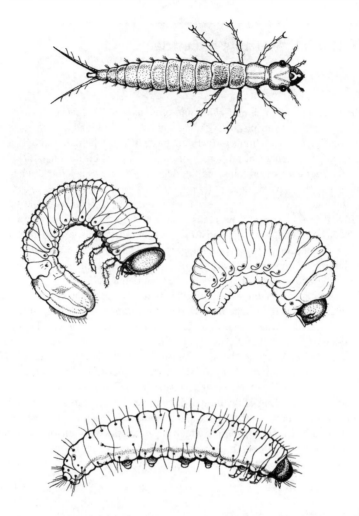

Figure 3.4 Examples of the shapes of beetle larvae. Reproduced from Munro (1966) with kind permission of Rentokil Initial plc

Small, hardened structures projecting from the end of the larval abdomen are called *urogomphi*. They are recognizable, for example, in the larvae of Dermestidae, Nitidulidae and Histeridae.

The third stage of metamorphosis is called the pupal stage. The pupa has mouthparts which do not articulate (i.e. are *adecticous*) and the rest of the pupal appendages are free and visible through the pupal coat (the pupa is *exarate*) (Figure 3.5). This is not so in the staphylinids, where the pupa is covered by a hardened coat and the pupal appendages are held in place by secreted material (an *obtect* pupa).

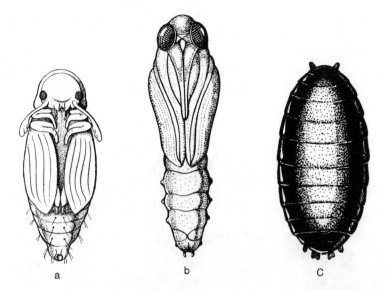

a   b   C

Figure 3.5 The types of pupae and puparia illustrate the relationship of appendages to the body. (a) The limbs are distinct from the body; this type of pupa is called an exarate pupa. (b) The limbs are firmly bound to the body; this form of pupa is called an obtect pupa. In the third type of pupa (c), the pupa is retained within the final larval coat; this coat is termed a puparium and the pupa is called a coarctate pupa. Reproduced from Munro (1966) with kind permission of Rentokil Initial plc

Some pupae pupate in a chamber within the soil. Others, like the scarabaeids, form a cocoon. In this instance the cocoon is made from material in the posterior section of the *caecum* (Richards and Davies, 1988).

The order Coleoptera is divided, on the basis of molecular studies, into what are treated as four suborders: Archostemata, Myxophaga, Adephaga and Polyphaga. The Archostemata is made up of three families which mostly inhabit decaying wood. The Myxophaga are made up of four families which are aquatic or are found in wet habitats and are algal feeders. Although all insects may be of importance in forensic entomology, the remaining two suborders, Adephaga and Polyphaga, contain families of beetles which are most commonly found at crime scenes. The suborder Adephaga contains 10 families and comprises predatory beetles which inhabit terrestrial and aquatic habitats and includes the ground beetles, the Carabidae. The Polyphaga contains 149 families including the families Dermestidae, Scarabaeidae and Staphylinidae.

3.1.1 Suborder Adephaga

These beetles are distinguished by the positioning of their legs. The coxae of the third pair of legs (the hind legs) are fused to the metasternum. When you look at the underside of the beetle you see that this region of the leg divides the first visible abdominal sternal plate (Figure 3.6).

There are lines down the sides of the thorax called sutures (the indents are positions where there is internal strengthening of the exoskeleton). An example of this is the suture between the notum and the sternum (sutures are readily recognized

Figure 3.6 The distinction between the Polyphaga (a) and the Adephaga (b), ventral view. In the Adephaga the hind coxa is fixed immovably to the metasternum, i.e. the coxa cannot move. It completely splits the first visible abdominal sternite. In contrast, in the Polyphaga the hind coxa is able to move, i.e. it articulates with the metasternum; usually does not generally divide the first visible sternite (A colour reproduction of this figure can be found in the colour section towards the centre of the book)

as the large transverse indentations across the thorax of the fly). The majority of beetles in this suborder have thread-like (*filiform*) antennae.

Larvae of insects in this suborder have legs with five segments which end in two claws (only rarely is it one claw). These larvae are mostly elongated and flattened (Luff, 2006; in Cooter and Barclay, 2006). Most beetles in the Adephaga are predaceous; in consequence they may feed on the insects inhabiting a cadaver.

3.1.2 Suborder Polyphaga

This suborder contains the majority of families of beetles with which the forensic entomologist may be concerned. The following features characterize this suborder. The hind coxa is rarely fused to the metasternum (it moves, or articulates) and so does not divide the first visible abdominal sternite. The thorax in this suborder does not have lines (sutures) across its dorsal surface. The types of antennae in the suborder vary, so they cannot be used as an indicative feature.

Polyphaga larvae are of many different shapes. They have legs with four segments which end in a claw. Some larvae in the suborder Polyphaga have legs which are reduced, others have vestigial legs, or the legs may even be absent altogether.

Polyphaga adults eat a variety of food. Some beetles are predaceous, but in the suborder as a whole many are phytophagous. Only beetles which are predators are of immediate importance to the forensic entomologist. A number of beetles visit a dead body, either because the body itself forms food and a habitat, e.g. the Dermestidae, or to feed on the insects already present, e.g. the Staphylinidae. The families of insects from this suborder that are important in forensic entomology include the Silphidae, Staphylinidae, Histeridae, Trogidae, Dermestidae. Cleridae and Nitidulidae.

3.2 Features used in identifying forensically important beetle families

3.2.1 Carrion beetles (Silphidae)

Silphidae have a flat body with sharp margins and their heads are small relative to the size of the thorax. The beetles of this family have antennae in which the sequence of antennal segments tends to thicken as the segments progress to the end, or the antennae are distinctly clubbed. The distance between the points of insertion of the antennae is wide. These are large, robust beetles and some, such as *Nicrophorus vespilloides* Herbst, have orange or red markings on their elytra. Others, such as *Nicrophorus humator* (Gleditsch) (Figure 3.7), are black in colour.

Figure 3.7   A silphid, *Nicrophorus humator* Gleditsch

One of the main identification features of this family is that abdominal segments protrude from the hardened upper wings (the elytra). If the beetle is turned over, six abdominal sternites are visible.

3.2.2   Rove beetles (Staphylinidae)

Staphylinidae are active beetles which are easily recognized because, when the insect is viewed from above (Figure 3.8), their short elytra expose at least half of the abdominal segments, so that seven to eight protrude. They range in size from tiny to large. For example the largest British staphylinid species *Ocypus olens* Müller (whose English common name is the devil's coach horse), has been recorded at 28 mm long (Richards and Davies, 1988). This family, however, are accomplished fliers and have strong membraneous wings packed away under their shortened elytra. Some species have the habit of curling up their last few abdominal segments over their 'back'. This makes them look very aggressive and the action is reminiscent of a scorpion. If you see specimens reacting like this as you approach them, then you are most likely seeing a staphylinid beetle.

Staphylinid beetles are predators and are attracted to the corpse to feed on the larvae of Diptera. A number of species of rove beetles (Staphylinidae) have

Figure 3.8   An example of the Staphylindae (A colour reproduction of this figure can be found in the colour section towards the centre of the book)

been found on a body; e.g. Goff and Flynn (1991) recorded the presence of adult *Philonthus longicornis* Stephens from a 23 year-old Caucasian male in Hawaii; and *Creophilus maxillosus* (Linnaeus), which Centeno *et al.* (2002) recognized as forensically relevant in their Argentine studies and which Chapman and Sankey (1955) also recorded from rabbits in exposed conditions in Surrey, UK.

3.2.3 Clown beetles (Histeridae)

These are small, shiny black beetles (Figure 3.9) with an exoskeleton that has a hard, often leathery or sculptured texture and a more or less oval shape. Their antennae are elbowed (geniculate) and the final segments of the antennae are formed into an obvious club. Histerid legs have flat tibiae. The significant identification feature of this family, when looked at from above, is the square-cut to the ends of the elytra, which reveal the last two abdominal segments.

Both larvae and adults are found on the corpse, as they feed on those insects attracted to decaying organic matter. The larvae also eat fly larvae and prey on other insects. The adult beetles respond to being handled by withdrawing their heads and pulling their legs, and any other projections, into the body, which is sculptured to allow this, and 'playing dead' (exhibiting *thanatosis*)

Figure 3.9 A hister beetle

3.2.4 Trogid beetles (Trogidae)

These are medium-sized beetles which are dull brownish in colour (Figure 3.10). The dorsal surface of the body appears roughened and the elytra can sometimes be hairy. The segments at the tip of the antennae are plate-like. The legs of trogid adults are not broad or modified for digging.

Trogidae larvae characteristically have long, sharp claws. Chinnery (1973) indicates that species of the genus *Trox* are not common in the UK. They are found at the dry stage on small carcasses and, in particular, feed on hide, fur, leather, feathers and dry matter. These beetles will also exhibit thanatosis if disturbed

3.2.5 Hide and skin beetles (Dermestidae)

Dermestidae range from very small to medium in size (1.5–10 mm) and have an oval to elongated shape (Figure 3.11). Their antennae are made up of 5–11 segments, ending in a club made of two or three segments (Peacock, 1993).

Adult members of the genus *Dermestes* lack a simple eye (an *ocellus*) on the head. The coxa on the front leg is conical and sticks out prominently from the coxal cavity (Figure 3.12). The femur of the hind leg is covered by the hind coxa, which is flattened into a plate. These beetles have the capacity to pull all their appendages into the underside of their body so that nothing protrudes.

Figure 3.10 · A trogid beetle

Figure 3.11 *Dermestes lardarius* (L.) (A colour reproduction of this figure can be found in the colour section towards the centre of the book)

Conical coxa

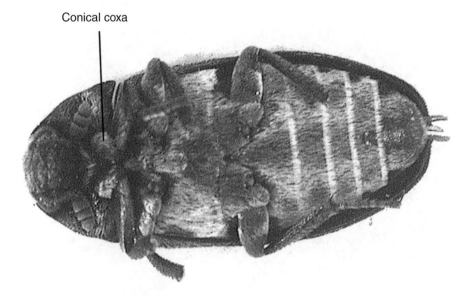

Figure 3.12   The front coxa of the dermestid projects from the coxal cavity (A colour reproduction of this figure can be found in the colour section towards the centre of the book)

Larvae of forensically-relevant Dermestidae are brown to black in colour and have hairs of varying lengths (*setae*) over the dorsal surface; there are frequently tufts of hair on the sides or posterior edge of the body. Indeed, the larvae of *Dermestes maculatus* DeGeer (Figure 3.13) are commonly known as 'woolly bears' as a result of this profusion of hairs. The larvae are 6–13 mm (1/4–3/8 inches) long and have two horns (urogomphi) on their terminal segment.

Dermestes lardarius Linnaeus is known to pupate in a puparium for 40–50 days at 18–20°C. They have one generation per year. Male *Dermestes lardarius* pass through four instars, whilst the female have five instars.

3.2.6   Checkered (or bone) beetles (Cleridae)

These beetles are usually brightly coloured on at least some part of their body (Figure 3.14). They are elongated and cylindrical in shape and appear to have a 'neck', because the first part of the thorax (the pronotum) is less broad than their elytra. The adults can be hairy. An example of a forensically significant member of the Cleridae is *Necrobia rufipes* DeGeer, the red-legged ham beetle, which can be found in association with bodies later in the decomposition sequence. In Hawaii it has been found in the soil under a corpse at a PMI of 34–36 days (Goff and Flynn, 1991). This species is a predator of fly larvae.

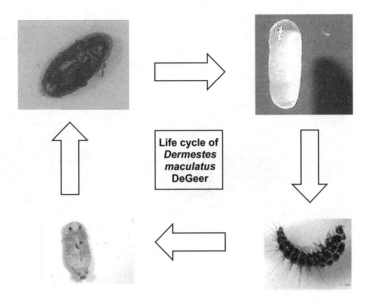

Figure 3.13 Life cycle of *Dermestes maculatus* DeGeer (A colour reproduction of this figure can be found in the colour section towards the centre of the book)

Figure 3.14 Clerid beetle (A colour reproduction of this figure can be found in the colour section towards the centre of the book)

3.2.7 Sap-feeding beetles (Nitidulidae)

These are very small beetles, and are not often longer than 7–8 mm (Figure 3.15). The Nitidulidae have recently undergone taxonomic revision. The antennae are usually

Figure 3.15   An example of a nitidulid beetle

composed of 11 segments, ending in a three-segmented club. The elytra are often truncated, but with rarely more than three abdominal segments visible dorsally. The fore- and mid-coxae are transversely orientated, whilst the hind-coxa is flattened. The tarsal formula for this family is most frequently 5–5–5 (this means that the tarsus of each of the legs is made up of five tarsomeres). The first segment (tarsomere) of the tarsus is not shortened and all of the tarsal segments are more or less dilated.

This family is a colonizer of corpses in the later stages of decomposition. According to Cooter and Barclay (2006), in the British Nitidulidae, the subfamily Nitidulinae includes two genera, *Nitidula* and *Omosita*, which are particularly associated with bones and dried carrion. Wolff *et al.* (2001) undertook a preliminary study in Medellín, Colombia, and found that 0.2% of the total number of families visiting a dead pig, which they had set up in an experimental 'crime scene', were members of the Nitidulidae. All members of this family were recorded from the advanced stage of decay which occurred 13–51 days after the pig died.

3.2.8   Ground beetles (Carabidae)

Ground beetles have a characteristic beetle shape. They can be found in a number of habitats, including grassland and forests. Carabids are members of the Adephaga because their first abdominal sternite segment is divided by the hind coxa. Their antennae are usually filiform, although some may be bead-like (moniliform), and are located on the head, between the eyes and jaws. The beetle head is prognathous.

Figure 3.16 Carabid beetle, illustrating the striations on the elytra

In carabids the elytra are usually sculptured, for example with striations, so that one sees nine regular ridges and furrows along the elytra (Figure 3.16). They are frequently fixed in position and, where this is the case, the beetle has only the vestiges of membranous wings.

Carabid larvae are long or elongated in shape. The larva has a pair of sharp pincer-like mandibles and six simple eyes (ocelli) down each side of the head. The larval abdomen has 10 segments and on segment nine there is a pair of *cerci*. The larvae have legs which end in two claws. Carabid larvae are very quick in their movements and tend to be nocturnal, so they may not be obvious members of the corpse assemblage.

3.3 Identification of beetle families using DNA

The same techniques of mtDNA analysis, use of RFLP and RAPD for the identification of flies, which are described in Chapter 2 (Section 2.3), are used in forensic entomology for the identification of beetles. Indeed, these techniques are also used in phylogenetic investigations of beetle species, such as that undertaken to separate members of morphologically similar ground (carabid) beetles of the *Nebria–Gregaria* group, on Queen Charlotte Islands in British Columbia, Canada. Clarke *et al.* (2001) concluded from RAPD and mtDNA analysis that only one species of the group could be separated out, on molecular grounds, from the particular group of carabids.

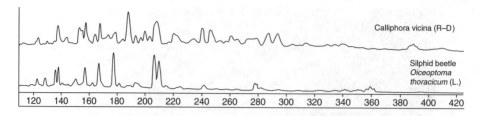

Figure 3.17 Profile of the RAPDs of a silphid beetle in comparison to a calliphorid fly. Reproduced with permission of Dr M. Benecke, reprinted from Benecke (1998), © 1998, with permission from Elsevier

RAPD analysis of beetle DNA has been a successful tool in crime analysis (Benecke, 1998). The families investigated included the carrion beetles (Silphidae), e.g. *Oiceoptoma thoracicum* Linnaeus, for which a DNA profile (Figure 3.17) was determined from a badly decayed body in October 1997 (Benecke, 1998). Mitochondrial DNA has also been used to identify the larvae of beetle species present on a body and also for additional purposes, such as identification of the human host from the gut contents of the larvae upon whom it had been feeding. Di Zinno *et al.* (2002) analysed specimens from the nitidulid genus *Omosita* in order to match mtDNA to a human host; this was successfully achieved.

Dobler and Muller (2000) explored the phylogenetic relationship of the Silphidae using 2094 base pairs (bp) of COI and COII, as well as tRNA. With the longer lengths of mtDNA, they were able to obtain a greater resolution of the genetic make-up of the family, providing an increased identification profile for use by the forensic entomologist. Zehner *et al.* (2004a) explored intra-species variation within the clerid beetles (based upon different mitochondrial genomes for the same species of organism – heteroplasmy). They showed that within the cytochrome oxidase I gene, in both *Necrobia rufipes* and *Necrobia ruficollis* Fabricius, there was a high degree of heteroplasmy which did not express itself as much in *Necrobia violacea* (Linnaeus), another species of clerid. This variation has to be considered when interpreting a profile from a specimen from the crime scene.

Less research has been undertaken on the molecular profiles of forensically important Coleoptera than for the Diptera. However, since the techniques are in place, further profiling of beetle species will expand this base as more crime scene investigations occur.

3.4   Further Reading

Benecke M. 1998. Random amplified polymorphic DNA (RAPD) typing of *Necrophorus* insects (Diptera: Coleoptera) in criminal forensic studies: validation and use in practice. *Forensic Science International* **98**: 157–168.

Benecke M. and Wells J. D. 2001. DNA techniques for forensic entomology. In Byrd J. H. and Castner J. L. (eds), *Forensic Entomology: The Utility of Arthropods in Legal Investigations*. CRC Press: Boca Raton, FL; pp 341–352.

Cooter J. and Barclay M. V. L. (eds). 2006. *A Coleopterist's Handbook*, 4th edn. Amateur Entomologists' Society: Orpington, Kent, UK.

Crowson R. A. 1981. *The Biology of the Coleoptera*. Academic Press: London.

Forsythe T. G. 1987. *Common Ground Beetles*. Naturalists' Handbooks No. 8. Richmond Publishing: Slough, UK; 74 pp.

Fowler W. W. 1887–1913. *The Coleoptera of the British Isles* (6 vols). Reeve: London.

Halstead D. G. H. 1963. Coleoptera: Histeroidea. *Handbooks for the Identification of British Insects* 4(10). Royal Entomological Society of London.

Hoy M. A. 1994. *Insect Molecular Genetics: An Introduction to Principles and Applications*. Academic Press: San Diego, CA.

Jessop L. 1987. Dung beetles and chafers (Coleoptera: Scarabaeoidea), 2nd edn. *Handbooks for the Identification of British Insects* 5(11): 2–33. Royal Entomological Society of London.

Joy N. H. 1976. *A Practical Handbook of British Beetles*. E. W. Classey: Faringdon, UK.

Lövei G. L. and Sunderland K. D. 1996. Ecology and behaviour of ground beetles (Coleoptera: Carabidae). *Annual Review of Entomology* 41: 231–256.

Simmons P. and Ellington G. W. 1925. The ham beetle *Necrobia rufipes* DeGeer. *Journal of Agricultural Research* 30(9): 845–863.

Tottenham C. E. 1954. Coleoptera: Staphylinidae, Section (a), Piestinae to Euaesthetinae. *Handbooks for the Identification of British Insects* 5(11): 2–33. Royal Entomological Society of London.

Unwin D. M. 1988. *A Key to the Families of British Coleoptera (and Strepsiptera)*. AIDGAP, Field Studies Council Publication No. 166. Headly Brothers Ltd: London [revised and reprinted from *Field Studies* 6(1): 149–197 (1984)].

Useful website

www.Coleoptera.org

4

The life cycles of flies and beetles

4.1 The life stages of the fly

Both flies and beetles have a life cycle which shows complete metamorphosis. This means that the different life cycle stages look dissimilar. The cycle starts when the adult female lays eggs and the following description relates to the life cycle of the fly.

4.1.1 The egg stage

Diptera tend to lay eggs in batches, and clumps of eggs are laid in places on the corpse that provide protection (Figure 4.1), moisture and food. In general, the number of eggs laid is around 150–200. Hinton (1981) suggests that *Calliphora vicina* may lay 2000–3000 eggs in their life time.

The blowfly egg is usually very shiny and white, ranging in size from around 0.9 mm to over 1.50 mm long and 0.3–0.4 mm wide (Rognes, 1991). The outer, textured coating of the egg is termed the *chorion*. This sculpturing which may, for example, be reticulate or spotty, can be used to identify different species of fly. If an electron microscope is available, investigating the surface of the egg may be a means of making identification to at least genus, of the fly species which has colonized the body (Greenberg and Singh, 1995). The end of the egg has a hole in it, called the *micropyle*. This is the route by which the sperm gain entry to fertilize the egg.

A furrow called the *plastron* runs the length of the egg along one side (Figure 4.2). This acts as a means of trapping air, should the egg become covered by water droplets or be drowned in water, and so aids continued respiration. The emergence of the first instar larva from the egg is called *eclosion*, although this term has also been used to describe any form of hatching. However, Erzinçlioğlu (1996) commented that he had experience of *Calliphora vicina* laying live larvae, where fertilization had taken place without a suitable oviposition site being immediately available.

4.1.2 The larval stage

The larva has 12 segments and a pointed anterior end, all that remains of the head capsule found in other insect larvae, with a black structure comprising mandibl

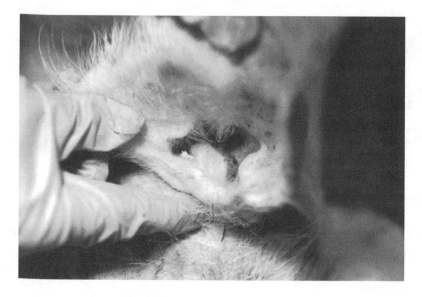

Figure 4.1   Clump of eggs laid inside the ear of a pig

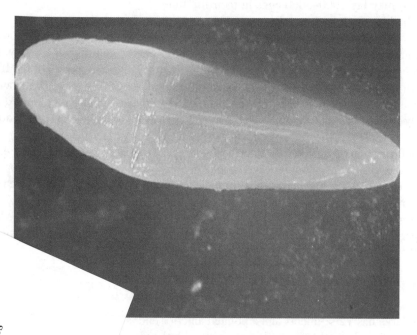

4.2   The plastron runs the length of the egg

and related sclerites and ending in mouth hooks (the cephalopharyngeal skeleton). The posterior end is blunt and has two brown circular areas on the final segment; these are the posterior spiracles.

In the fly there are three larval stages or instars (see Figure 4.3) and a particular larval stage out of the three is distinguished by referring to it as L1, L2 or L3 (or sometimes LI, LII and LIII). The specific life stage of the larva can be identified by the number of slits present in each posterior spiracle. In the first instar one slit is present; in the second instar two slits are present; in the third instar three slits are present. In blowflies there is normally a difference in size of larvae in the three larval stages; the first instar tends to be less than 2 mm in length, whilst the second instar is 2–9 mm long and the third instar can be 9–22 mm long. However, size is a relatively unreliable measure of age because it is dependent upon the amount and quality of food available– although a body may be considered to be an abundant source of food.

Projections called tubercles (Figure 4.4) surround the edge of the posterior segment of the larva. The spiracles are located on the horizontal face of this final

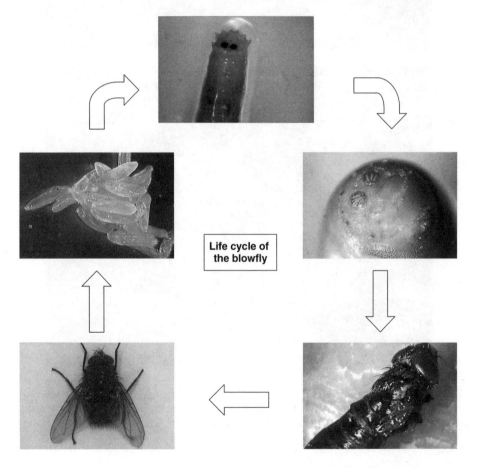

Figure 4.3 Life cycle of the blowfly (A colour reproduction of this figure can be found in the colour section towards the centre of the book)

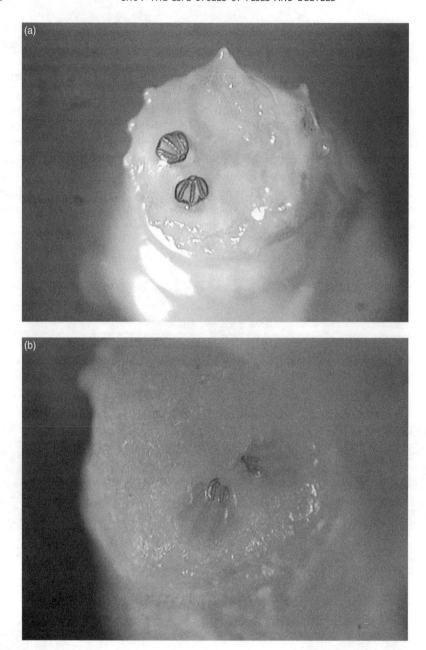

Figure 4.4 (a) Posterior section of the larva, showing tubercles and posterior spiracles. (b) Transition from larval stage 2 to larval stage 3, showing emerging spiracle slits (A colour reproduction of this figure can be found in the colour section towards the centre of the book)

posterior segment. The distance between tubercles plays a role in the identification of larval species. For example, in larvae of *Lucilia sericata*, Smith (1986) points out that the inner tubercles (those at 12 o'clock) are separated from each other by a distance roughly equal to the distance between the inner and the median tubercles (those at 10 o'clock and 2 o'clock, respectively; see Figure 4.4).

Sticking out from the third anterior segment (second thoracic segment) of the larva there is an anterior spiracle, which looks like a hand with fingers projecting from it (Figure 4.5). The morphology of this spiracle can also be used as a means of identification in some species.

Larvae in the third instar are the largest, and half-way through this stage they stop feeding and become migratory, seeking a place for *pupariation* (the final developmental stage of metamorphosis into the adult stage). This is called the larval post-feeding stage. Larvae move away from the body, towards dark and somewhat cooler areas. In the post-feeding stage the contents of the crop begin to reduce, until finally there is no obvious dark line of crop material visible through the white larval cuticle. Cragg (1955) suggests that the post-feeding larvae may move up to 6.4 metres from the carcass. On concrete floors, such as might be found in *buildings*, post-feeding larvae have been known to migrate up to 30 metres from the body (Green, 1951). Usually the post-feeding larva attempts to bury itself in soil or some other dark location. They may be found by searching in the first 2–3 cm depth of soil at

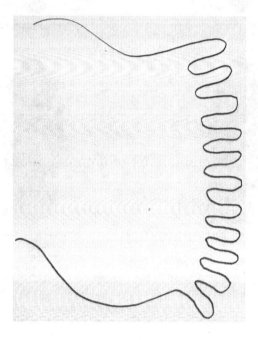

Figure 4.5 Example of the anterior spiracle of the larva. The shapes can vary and can be used for identification purposes

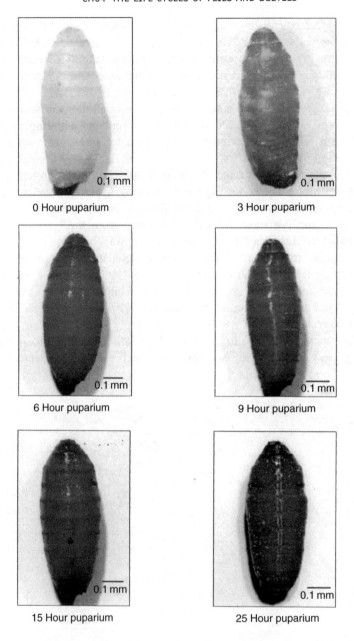

0 Hour puparium   3 Hour puparium

6 Hour puparium   9 Hour puparium

15 Hour puparium   25 Hour puparium

Figure 4.6   Puparial colour changes of *Calliphora vomitoria* (L.) up to 25 hours after the onset of pupariation (A colour reproduction of this figure can be found in the colour section towards the centre of the book)

outdoor crime scenes. This tendency to migrate is not true for all species; some, e.g. *Protophormia terraenovae*, have been found to pupate on the corpse (Erzinçlioğlu, 1996).

The puparial case changes colour over time, becoming an oval object resembling an uncut cigar, coloured somewhere between reddish-brown and a dark mahogany brown or black (Figure 4.6). This case maintains all of the features of the third instar, so there is some possibility of identifying this stage to species, using keys for the identification of third instar dipteran larvae. Some attempts have been made to relate the state of colouration development of the puparium to post mortem interval, but to date the methods have not shown great accuracy beyond the first 24 hours (Greenberg, 1991).

Emergence of the adult, at the end of the life cycle, is achieved by its pushing the cap (*operculum*) off the puparium, using a blood-inflated region on the head called a ptilinum. This is like an 'airbag' which projects from the anterior-dorsal region of the head as the fly emerges (Figure 4.7). It later sinks back into the facial structure, generating the crease, or ptilinal suture, just above the antennae. The mouth hooks (cephalopharyngeal skeleton) remain inside the broken puparial case and can be used to confirm identification if you can find them.

The adult pushes out of the puparial case and up through the soil, responding, according to Fraenkel (1935), to light intensity. Once above the soil surface, the fly evacuates the waste products of pupation as a greenish-black liquid. This material is called the meconium (the same term is used for the first stool of the human infant).

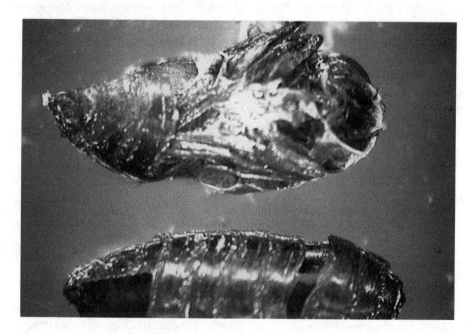

Figure 4.7 Adult fly emerging from the puparial case (A colour reproduction of this figure can be found in the colour section towards the centre of the book)

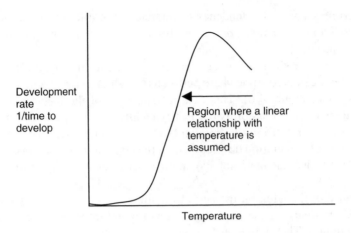

Figure 4.8 Generalized insect growth curve

The fly 'dries out' and eventually its wings expand and the greyish-coloured fly becomes recognizably pigmented as, for example, a bluebottle or greenbottle.

The speed of insect development is determined by temperature (Figure 4.8). A number of researchers in a number of countries have determined the duration of the life stages of locally relevant species of fly, at particular temperatures. These include Kamal (1958), Greenberg (1991), Reiter (1984), Anderson (2000) and Grassberger and Reiter (2002). This information forms the experimental measure for post mortem interval, because from it the energy budget for development (the accumulated degree hours) can be determined. It is important to use the sum of the duration of each of the individual life stages and to work with both the average and the maximum and minimum durations, at the specific experimental temperatures used. Further discussion of the method for calculating the post mortem interval is found in Chapter 7.

4.2 The life stages of the beetle

Beetles are insects which also show complete metamorphosis. The term for this is *holometabolous*. To become adults, they too pass through an egg stage, three to five larval stages depending on species, and a pupal stage. Coleopteran eggs tend to be oval, spherical or spheroid in shape and are usually considered very similar, irrespective of family. Beetles usually bury themselves in the ground, or in a specially constructed chamber, when they pupate. Less detailed information is available about beetle life cycles than is known about the Diptera.

The length of the life cycle will vary, depending upon the family and species of beetle. Development through the complete life cycle, from egg to adult (imago) can take 7–10 days in rove beetles (Staphylindae), whereas in the ground beetles (Carabidae) completion of the life cycle to the adult stage may take a year and the adults may live for 2–3 years. In some species the number of instars in the larval stage is not fixed, but is dependent on environmental conditions. In Dermestidae, for example, there may be as many as nine instars (Hinton, 1945). Usually, however, there is only one generation of beetles per year. Smith (1986) indicates that the length of the pupal stage of *Dermestes* sp. can last between 2 weeks and 2 months and that these beetles can overwinter (enter *diapause*) in a pupal chamber if the weather is not suitable or it is late in the season.

The problem of lack of a ready morphological distinction between larval instars in beetles means that other methods to distinguish the instar are needed. Watson and Carlton (2005) investigated the life cycles of three species of American silphid which feed on both the insects visiting the carcass and on the carcass itself; *Oiceoptoma inaequale* (Fabricius), *Necrophilia americana* (Linnaeus) and *Necrodes surinamensis* (Fabricius). Using multivariate analysis, they identified three larval stages. For *Necrodes suranamensis* they showed an average duration of 12 days for the first instar, 10 days for the second instar and 11 days for the third instar, indicating that the larvae were present on days 9–22 of decomposition (Watson and Carlton, 2005).

Chapter 9 provides further information about some of the beetle families which play a significant role in succession on a corpse and so have a role in forensic entomology.

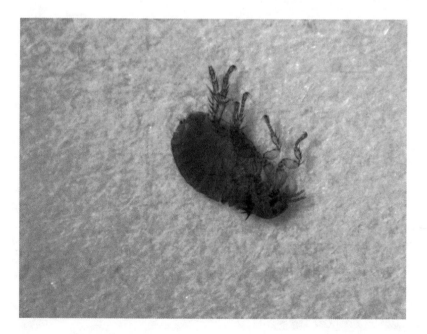

Figure 4.9 A flea, an ectoparasite which may be alive on a submerged body for up to 24 hours

4.3 The influence of the environment on specific insect species

In interpreting evidence based on forensic entomology, it is important to consider both the ecology of the insects (discussed in Chapters 8 and 9), as well as the information about the crime scene that can be deduced from the presence of these insects. For example, whether the insects are alive or dead is important in interpreting the history of the body after death. Some people may have ectoparasites, such as fleas (Siphonaptera) and lice (Phthiraptera), on them when they are alive. Whether these are found dead or alive on the corpse is important as, in the case of fleas, is their recovery rate, if the body has been immersed in water (Simpson, 1985). According to Simpson, fleas (Figure 4.9) present on a body will drown in 24 hours if the body is submerged; however, if recovered from the body, fleas can revive after 60 minutes if the immersion was for up to 12 hours. A period of 4–5 hours will be needed for the fleas to revive if the body has been submerged for 18–20 hours. Lice (Figure 4.10) are less robust and will die if the body has been submerged for only 12 hours. Brisard (1939; cited in Smith, 1986) found that lice remain on a recently dead body which is on dry land, for as long as the body temperature is suitable. Working at 26°C and 65 % RH, Durden (2002) recorded a demise of 4 % in a population of human lice, 24 hours after they had been removed from a live host. He found that at 22°C most lice die within 7 days.

Figure 4.10 The human louse, an indicator of recent death if still alive on a corpse

Blowflies respond to immersion even more quickly than do fleas and lice. If the body is submerged in water, having previously had eggs laid on it which have hatched, then information on the place of death can be interpreted from the state of the larvae on the body. For example, if the larvae are alive, according to Arhuzhonov (1963; cited in Smith, 1986), the body has been recently moved.

Dry conditions at a crime scene can influence the nature of the corpse, which frequently dries out if death occurs in a centrally heated house or the body is concealed in a dry location. If a body is bricked up in a chimney, incarcerated in cement, such as in the foundations of houses or bridges, or has been placed under the floor boards, then clothes moths, mites and dermestid beetles will be the dominant colonizers.

If disposal is by burning, whether the burnt corpse supports colonizing insects will depend on the degree of burning to which the corpse has been exposed. Avila and Goff (1998) suggested that corpses burned with petrol to second-degree burn condition were more attractive than unburned corpses. They found that calliphorids [*Chrysomya megacephala, Chrysomya rufifacies, Lucilia cuprina* (Weidmann) and *Lucilia sericata*] colonized a burnt pig 1 day earlier than an unburned one. Meek (1990) showed that car boots (trunks) containing burnt bodies retarded the attraction of the flesh by around a week, because of the lack of attractiveness of the body surface. Joiner (2003, unpublished) demonstrated that *Lucilia sericata* was able to oviposit in cracks on flesh burnt to the degree described as stage 2 in the Glassman and Crow (1996) scale. The larvae developed successfully. However, gravid females showed interest in, but did not land upon, flesh burned to stage 4 of this scale. So it is the degree of burning and the accessibility of unburned or slightly burned flesh which is significant in determining the level of blowfly colonization, rather than solely whether or not the body has been burned. Indeed, where the corpse had decomposed, been colonized by insects and then subsequently been burned, Anderson (2005) demonstrated that the forensic evidence could survive sufficiently for the time since death to be determined.

The effect of darkness and low temperature on insect colonization has also to be considered in interpreting time since death. Crime scenes have been known where the insect-infested bodies have been covered in snow (Wyss *et al.*, 2003a); and bodies have been removed from caves in Switzerland, where they have decomposed in conditions of apparent total darkness at 5°C and yet have been infested with insects (Faucherre *et al.*, 1999). The ecological conditions associated with the activities of flies are discussed further in Chapter 8. Environmental conditions may affect not only the presence, but also the genetic make-up of the fly. Work by Zhong *et al.* (2002) has shown that very low magnetic fields have an effect upon messenger RNA and its transcription of cytochrome oxidase. They suggest that there is a biological effect from low-level magnetic fields, even where exposure is as low as 20 minutes. Therefore, it is important to consider which species will present at a crime scene, as a result of the prevailing environmental conditions, and also to consider the influence of conditions, including non standard aspects such as magnetic fields, on the particular crime scene analysis which is being undertaken.

Where there has been a knifing or a shooting and the body is not immediately found, blowflies may visit the body and spread droplets of the blood around the crime scene. This can cause confusion in interpreting the blood spatter pattern because of the tendency of blowflies to feed on areas oozing blood. Flies can generate small spots of blood by regurgitation, defaecation and also because they walk in fresh blood and move the blood to another location. Benecke and Barksdale (2003) identified the features of blood spatter which were characteristic of the activity of flies. They concluded that the spots would resemble a tadpole, or be sperm-like in shape (such stains, which resembled spermatozoa, 'did not end in a small dot'), and that the stains would have a tail-to-body ratio of greater than one. To prevent confusion between genuine blood spatter and the activities of flies, they suggested excluding any stains that did not have a distinguishable tail or head, stains with an irregular structure and stains which do not show directionality, from the blood spatter analysis. Larger blowfly stains will have randomness in their distribution, whereas blood spatter of higher animal origin, where there is blood pressure to cause the spatter, will be grouped so that a convergence point can be deduced.

4.4 Succession of insect species on the corpse and its role in post mortem estimation

The succession of the waves of insects colonizing a body have been studied by a number of researchers, including Mégnin (1894); Hough (1897); Chapman and Sankey (1955); Bornemissza (1957); Payne (1965); Easton and Smith (1970); Lane (1975); Rodriguez and Bass (1983); Goff (1993), who provided a summary table of decomposition studies until 1993; Dillon and Anderson (1996); Carvalho *et al.* (2000); and Centeno *et al.* (2002). As Goff (1993) points out, some of the insects present on the corpse are only reflective of that location or habitat, whilst others are present on the corpse because of a direct relationship to stage of corpse decomposition and so can be used for determining the post mortem interval. The sequence of colonization by insects is different, depending upon whether the body is buried, left on the soil surface or has been thrown into water.

4.4.1 Succession on a buried corpse

The diversity of fauna colonizing buried corpses is smaller than for those colonizing a corpse exposed above ground. Payne *et al.* (1968), using pigs buried at depths of 50–100 cm, listed 48 arthropod species colonizing the body; 26 of these species were restricted to buried corpses. Six to eight weeks after burial of a pig carcass, they recorded 80 % decomposition, based on weight loss. This was in contrast to decomposition to the same stage which, for unburied pig carcasses, took place in 7 days. Other workers have found that blowflies tend to be absent from buried corpses, although muscids can be found on a corpse covered by 2.5–10 cm of soil (Lundt, 1964).

Bodies buried at a crime scene may be revealed by the presence of coffin flies, *Conicera tibialis*. Their presence on the soil surface could indicate the location of a body buried for at least a year and possibly longer. Colyer (1954) found phorids on a patch of soil below which, some 12 months earlier, the family pet had been buried. Being an enthusiastic entomologist, he dug up the pet and confirmed that the coffin flies were indeed originating from the body of the dog. Colyer also linked *Conicera tibialis* [misnamed as *Gymnoptera vitripennis* (Meigen)] specimens, submitted to a museum, to the presence and duration of burial of a human body. The phorids were recovered as puparial cases on hair from a corpse that had been exhumed, after being buried at Hay, Herefordshire, for 10 months, confirming again that this family can be considered as indicators of burial for long periods of time. Further research, however, showed that this was not a response whose timing could be predicted on an annual basis, as the flies can pass through a number of generations below ground before emerging. So they are an indicator of buried bodies, but not necessarily of the duration of the burial.

A number of beetles will dig down through the soil to gain access to a corpse. These include members of the Staphylinidae and Rhizophagidae. Hakbijl (2000) relates the presence of *Rhizophagus paralellocollis* (Gyllenhal) as inhabiting corpses buried for 2 or more years, and feeding on the fungi present on the corpses. He notes that this species is regularly associated with nineteenth century burials. Payne *et al.* (1968) related species to the state of decomposition in buried corpses and these are presented in Table 4.1.

Table 4.1   Succession of insects on buried bodies

Common name	Family
Mites	Uropodidae
Acarid mites	Acaridae
Scuttleflies	Phoridae
	Scatopsidae
Gall midges	Cecidomyiidae
Fruit flies	Drosophilidae
Gall wasps	Cynipidae
Diapriid wasps	Diapriidae
Milichiid flies	Milichiidae
Lesser dung flies	Sphaeroceridae
Ground beetles	Carabidae
Springtails	Collembola

After Payne *et al.* (1968).

4.4.2 Succession on a corpse above ground

No specific insect has been associated with a particular species of corpse, according to Hough (1897), who worked on bodies of horses, snakes, cats, dogs, fish and humans which were left exposed on the surface of the ground. Despite this lack of specific relationship, because the pig is considered biologically similar to the human, it is the preferred model for forensic entomology research on succession in humans, where human cadavers are not used or for legal reasons cannot be used.

Successional changes in insect fauna on the corpse have been noted and related to the stages of decomposition through which the body passes. However, as described in Chapter 1, changes in the insect species present can be the result of seasonal changes and annual conditions, as was demonstrated in Australia (Archer and Elgar, 2003). A summary of the seasonal difference in insect species of European

Table 4.2 Seasonal distribution of European insect species on a body

Family	Genus	Season
Calliphoridae	*Calliphora*	Spring, autumn
	Lucilia	Summer
	Phormia	Summer
	Pollenia	Summer
	Melinda	Summer
	Cynomya	Summer
Sarcophagidae		Summer
Rhinophoridae		Summer
Tachinidae		Summer
Fanniidae	*Fannia*	Spring, summer, autumn
Muscidae	*Azelia*	Summer
	Hydrotaea (Ophyra)	Spring, autumn
	Muscina	Summer
	Musca	Summer
	Pyrellia	Summer
Anthomyiidae	*Phorbia*	Summer
	Paregle	Summer
	Anthomyia	Summer
	Fucellia	Summer
Platystomatidae		Summer
Dryomyzidae		Summer, autumn
Heleomyzidae		Summer, autumn
Piophilidae		Spring, summer, autumn
Carnidae	*Meoneura*	Spring, summer
Milichiidae		Spring, summer
Agromyzidae		Spring, autumn
Chloropidae		Summer
Clusiidae		Spring

Modified from Dear (1978), with permission of the Amateur Entomologists' Society.

origin recorded generally on carrion, although not specifically on a human corpse, was provided by Dear (1978). The list is presented in Table 4.2. As is clear, the proportion of colonizers is most rich in the summer months.

Gaudry *et al.* (2004) assessed 400 or so forensic cases in France, and showed that Sarcophagidae appear in the first wave of colonizers on the unburied body. However, they pointed out that the family arrived a little later in the first wave of colonization than did *Calliphora* and *Lucilia* spp. Muscid flies such as *Musca domestica* will also visit a corpse soon after death, not in order to feed upon the corpse but to consume any exudates, such as faeces or urine, that have been voided. In Hawaii, the muscid fly *Synthesiomyia nudiseta* (van der Wulp) is a significant initial colonizer species and an indicator of urban locations, being absent from locations in rural areas (Goff, 2000). It has also been used in South Carolina, USA, as a means of determining post mortem interval, although it was a newly recorded immigrant (Lord *et al.*, 1992). The life cycle for *Synthesiomyia nudiseta* has been confirmed at two temperatures in the laboratory for specimens originating in Venezuela; at 20°C it takes 27 days to complete its life cycle and at 28°C this is reduced to 17.8 days (Rabinovich, 1970).

4.5 Review technique: preparing slides of larval spiracles or mouthparts – preparation of whole slide mounts

4.5.1 Introduction

On some occasions it may be appropriate to make a microscope slide of a specimen to confirm aspects of its identity. To do this you will need to 'clear' a specimen, so that you can examine its spiracles more easily. Soft tissue is removed and the chitinized parts of the specimen remain, so that they can be mounted on a microscope slide. The same procedure can be used to make a slide of the puparium, in order to attempt to identify the species to which a puparium belongs, or to prepare a slide of the cephalopharyngeal skeleton.

The procedure described below is for the preparation of a slide of the spiracles of a larva to confirm its instar.

4.5.2 Safety instructions (Control of Substances Hazardous to Health – COSHH)

- 10 % KOH is CAUSTIC – wear latex gloves.

- Glacial acetic acid can cause irritation – keep the watch glass covered.

- Use normal laboratory etiquette – wear safety glasses for preparing the slide.

4.5.3　Materials

2 Dropper pipettes
3 Watch glasses
Slides
Cover slips
Labels for slides
Wash bottles with distilled or deionized water
Glacial acetic acid
Compound microscopes with phase contrast facility
Beaker for waste
10 % Potassium hydroxide
Mounted needles
10 ml Measuring cylinder
250 ml Beaker
Mountant, e.g. Euparol
Fine forceps
Specimen of larva (or puparium)
Glass rod
Clove oil

4.5.4　Method

1. Immerse your specimen (which you have carefully punctured with a fine pin), in 10 % potassium hydroxide (KOH) overnight to soften the tissue and destroy the internal tissues (the alternative is to boil the specimen in 10 % KOH, but this is a potentially dangerous procedure as the potash solution (KOH) is caustic).

2. Place the whole specimen in a small watch glass and flood it with distilled water to wash off the potassium hydroxide.

3. Pipette off the liquid and dispose of it in the waste beaker.

4. Repeat the washing in distilled water as in instructions 2 and 3. Your specimen should now be either a very pale yellow colour or transparent (if not, replace it in the potassium hydroxide for a longer period, until you are satisfied with the level of transparency of the cuticle).

5. Remove all water and blot the specimen dry. *CARE!* The specimen must be dry for the next stage.

6. Using fine forceps place the specimen into a second, *DRY* watch glass and add 5 ml glacial acetic acid with a fresh dropper pipette. Cover the dish to reduce the rising odour level (ideally this stage should be carried out in a fume cupboard).

7. Leave the contents for 5 minutes to dehydrate your specimen.

8. Using forceps, place the specimen into a third watch glass and add (by eye) 5 ml of clove oil. Cover the watch glass with two glass slides to reduce the odour (again, this should be carried out in the fume cupboard).

9. Leave the specimen for 10 minutes. Meanwhile, set up the compound microscope and collect the slides and cover slips.

10. Add a small amount of mountant (e.g. Euparol) to the centre of a glass slide, using a glass rod.

11. Take the specimen out of the clove oil. Position it carefully on the glass slide, using mounted needles.

12. Carefully overlay the specimen by sliding a cover slip down a mounted needle onto the specimen. Label the slide with your name, the date and the contents of the slide.

13. Allow the slide to dry and harden. This can take 2 weeks (or use a slide dryer to reduce the time).

14. Examine the specimen under phase contrast ($\times100$). Make a labelled diagram of the spiracles and indicate the instar of the specimen in your laboratory notebook.

15. Justify your conclusion and use Kamal's data to determine the duration of the life cycle of your specimen to this point if grown at 26.7°C.

4.6   Further reading

Carvalho L. M. L. and Linhares A. X. 2001. Seasonality of insect succession and pig carcass decomposition in a natural forest area in south eastern Brazil. *Journal of Forensic Sciences* **46**(3): 604–608.

Erzinçlioğlu Y. Z. 1989. The value of chorionic structure and size in the diagnosis of blowfly eggs. *Medical and Veterinary Entomology* **3**: 281–285.

Erzinçlioğlu Y. Z. 1985. The entomological investigation of a concealed corpse. *Medicine Science and the Law* **25**(3): 229–230.

Grassberger M. and Reiter C. 2002. Effect of temperature on development of the forensically important holarctic blow fly *Protophormia terraenovae* (Robineau-Desvoidy) (Diptera: Calliphoridae). *Forensic Science International* **128**: 177–182.

Greenberg B. and Singh D. 1995. Species identification of calliphorid (Diptera) eggs. *Journal of Medical Entomology* **32**: 21–26.

Harvey M. L., Dadour I. R. and Gaudieri S. 2003. Mitochondrial DNA cytochrome oxidase I gene: potential for distinction between immature stages of some forensically important fly species (Diptera) in western Australia. *Forensic Science International* **131**: 134–139.

Hinton H. E. 1960. Plastron respiration in the eggs of blowflies. *Journal of Insect Physiology* **4**(2): 176–180.

Lefebvre F. and Pasquerault T. 2004. Temperature-dependent development of *Ophyra aenescens* (Weidemann, 1830) and *Ophyra capensis* (Weidemann, 1818). *Forensic Science International* **139**: 75–79.

Nuorteva P. 1987. Empty puparia of *Phormia terranovae* R-D (Diptera: Calliphoridae) as forensic indicators. *Annales Entomologici Fennici* **53**: 53–56.

Pont A. C. 1979. Sepsidae: Diptera, Cyclorrhapha, Acalyptrata. *Handbooks for the Identification of British Insects* **X**(5c). Royal Entomological Society of London.

Rozen J. G. and Eickwort G. C. 1997. The entomological evidence. *Journal of Forensic Sciences* **42**: 394–397.

5

Sampling at the crime scene

The most efficient way to respond to an invitation to attend a crime scene is to have ready a carrying case containing all of the requirements for collecting entomological specimens. This means that there is no great delay in reacting to a request and most contingencies can be addressed. If you are asked to attend the crime scene by the police, you need to collect sufficient samples so that material can be made available, if necessary and requested, for a fellow forensic entomologist to make his/her own assessment for the defence. This is imperative, since if the body is buried or cremated and the relatives wish to view it prior to disposal, they require the body to be insect-free (it may also be that those instructing you in other contexts need to undertake hygiene control activities, which mean that the environment concerned will change; so speed and good sampling techniques are required in this situation).

5.1 Entomological equipment needed to sample from a corpse

The entomological equipment (Figure 5.1) required includes plastic or polycarbonate screw-top sampling jars for both preserved specimens and live cultures, forceps, stepping plates to preserve the scene from contamination, a killing jar containing ethyl acetate, labels, indelible markers with fine points, fine forceps, artists' paint brushes, an entomological net and killing agents for larvae, such as boiling water, and an insect preservative. A number of preservatives could be used, including 70–80 % alcohol, KAAD and Kahle's solution; each has its benefits.

Kahle's solution contains both a fungal control and a preservative. It has been used at the University of Lincoln for 5 years and has preserved the samples used in a teaching collection in the same flexible condition as they were when the larvae were first killed. Alcohol has also been used, but this has required that the samples were more frequently curated than when using Kahle's solution, because of evaporation.

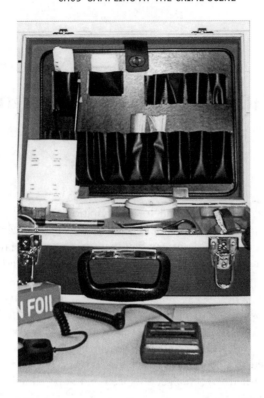

Figure 5.1 The contents of an entomological scene-of-crime case, with equipment

Kahle's solution can also be used to kill larvae if all else fails, although this is not a recommended approach. It is a preservative for dead adult insects, and so provides a means of combining uses and limiting the amount of equipment and chemicals required at the scene. References in the Further Reading section provide details of the effects on size of several preservatives and indicate why it is valuable to kill the larvae using boiling water, or water at a temperature just less than boiling (Adams and Hall, 2003).

Because live specimens must be recovered from the site, it is necessary to bring some food for them. Liver, such as pig's liver, or minced (ground) beef has been found to be the most suitable (although it should be noted that research indicates that larvae show variable growth on different body parts). The food should ideally be at room temperature, not frozen or chilled, when the maggots are placed upon it. For the return journey the cultures should be kept in as low a temperature as possible, ideally below the base temperature of the specimens. A mobile refrigerator for the car or van, or a cool box with artificial ice blocks, would be ideal. A thermometer should be included in the container to ensure that the temperature during transport can be confirmed.

A carrying box, or packaging for the specimens, should be included. The sample jars of preserved and live specimens, from each site on the body, should be retained

---

Box 5.1 Composition of Kahle's solution[*]

95% Ethyl alcohol[**]	30.0 ml
Formaldehyde[#]	12.0 ml
Glacial acetic acid[##]	4.0 ml
Water	60.0 ml

[*]Care should be exercised in the storage of these chemicals.
[**]Ethyl alcohol is flammable.
[#]Formaldehyde is toxic.
[##]Glacial acetic acid is CORROSIVE; the acid should be added slowly to water and NOT the other way round.

---

together in pairs. Where samples are being taken by a crime scene investigator (scene of crime officer, or SOCO) rather than the forensic entomologist, it is necessary to package the samples and seal them, so that the integrity of the sampling is not at risk. These storage packages can be individual cardboard boxes, which are sealed with both preserved and culture samples from the same site on the body, in the same package. In this instance, the package requires to have holes punched in it and the lids to the culture jars need also to have holes, or a porous covering, which is firmly attached to the top of the container. Larvae are 'escape artists' and will push through a top if it is not secured. If this happens, your evidence will have escaped! The French Gendarmerie use polythene bags which are appropriately labelled and sealed as their means of packaging at a crime scene (Figure 5.2). Pin holes are made through the bag to prevent a build-up of carbon dioxide, whilst preventing the larvae from escaping.

In order to kill larvae from each colonization site on the body, they are immersed for 30 seconds in recently boiling water, to fix the larvae at their maximum length. Water can be brought to the crime scene in a thermos flask, or prepared on-site using a small camping stove and kettle (matches or a gas lighter are also required if you are boiling water on site!).

The general habitat at the crime scene should be recorded. This includes: whether the body has been wrapped or covered in some way (Figure 5.3); if indoors, whether the windows are open or closed; the slope of the ground if the crime scene, or where the body was found, is outside; the nature of any vegetation and a general site description, along with associated photographs, should be recorded. The crime scene temperature must also be recorded, along with the degree of light or shading at the scene.

Thermometers should be included in your equipment case. These thermometers should be calibrated so that they read accurately and do not give readings which have to be corrected. For safety reasons, if a digital probe thermometer is not used, it is better to use an alcohol thermometer rather than a mercury thermometer. The

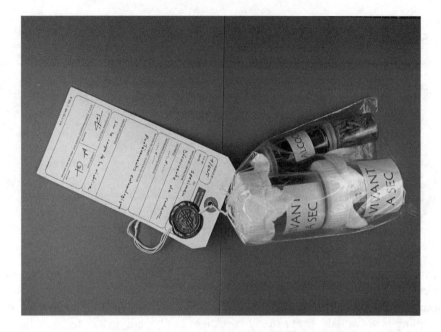

Figure 5.2 Sealed, labelled bag containing specimens of insects collected from the crime scene. Reproduced with kind permission of the Institut de Recherche Criminelle of the French National Gendarmerie

Figure 5.3 Body wrapped in black polythene

thermostat should be noted on any central heating units which operate indoors and which might dictate the conditions in the building. If at all possible, a weather recorder should also be brought on site, if it is an outdoor location, so that the temperature, light intensity, humidity and wind direction and speed can all be recorded over a period of time. At the minimum, a temperature and humidity recorder should be used on site.

5.2 The sampling strategy for eggs

Once permission from the senior investigating officer has been obtained, the body should be searched in an orderly sequence. The head region is examined first and then the trunk is searched, moving along towards the legs and toes, which are separated and checked. Any wounds are specifically noted. Once one side has been checked, the body should be turned over and the under-side should be examined. Clothing can be examined cursorily on site. In particular, the pockets, sleeves and clothing folds can be checked at the scene with the agreement of the officer in charge. A more thorough search is possible at the mortuary when the clothes, if present, are stripped from the body. Fly eggs are laid in batches and can be mistaken for everything from yellowish white mould to sawdust, or an encrusting of salt on the body; beetle eggs are often laid individually, so may be easily missed at the crime scene.

Fly eggs are normally laid in or near dark, moist orifices of the body, such as the ears, nose, eye lids, mouth or genitalia. They may also be laid in folds of skin behind the ears, in joint creases, or on clothing which has absorbed body fluid exudates. So it is important that all sides of the body are examined and it may be necessary to attend the post mortem to check further for insects, if the body is fully clothed or has been wrapped in something. The individual clumps of eggs should be picked off and carefully placed in a container without any food. The humidity in the container can be maintained by using a damp paper towel placed in the tube to stop the eggs drying out.

Each sample should be given an item number and the crime scene details. The label should be written in indelible ink (not ball point ink, as this will not survive damp conditions). The label should include the name of the crime scene investigator who collected the sample (Figure 5.4), the officer in charge of the

```
Crime Scene No

Officer in charge

Collector

Date

Item No.

Location and description
```

Figure 5.4 A label containing information about the scene, date, collector, the crime number and item number should be included inside and outside the collecting jars used at a crime scene

case, the case number, the item number, the date and location on the body from which the sample was taken. This label should be placed on the body of the container, whilst a non-adhesive version is placed inside the container.

Placing this information on a label both within the container and outside it limits the likelihood of losing the information and ending up with a sample of unknown origin. The easiest way of getting a paper label into a container is to roll it round a pencil or paintbrush handle and deposit the roll through the neck of the container, where it unrolls. These data must also be recorded in your scene log.

5.2.1 Larvae

Larvae will be located as the body is searched for eggs. They too tend to be in the orifices, such as the eyes, ears, nose and so on, including any wounds which were made on the body. The larvae should be collected from each site in batches of 20–30 per jar, so that no additional heat or ammonia is generated during transit. More than one collection jar per infestation site may be needed. The first instar is the smallest and most vulnerable of the three larval stages and the larvae, if sampled at this stage, can easily die. It is necessary, therefore, to protect them from drying out when collecting and culturing these from a corpse at a crime scene.

Boiling water is poured into a container such as a styrene cup or a collecting jar to a depth of 3 cm, and larvae which are to be preserved from the specific site are then added. They are left immersed in the water for at least 30 seconds before the contents of the jar are poured through a small sieve and collected in a large waste container. Large bottled water or catering fruit juice containers make excellent waste containers. The contents of the waste container, when full, can be poured down a foul sewer or toilet, away from the crime scene.

Larvae are known, when they reach late second and third instar stages, to mass together. These maggot masses are capable of raising the temperature above ambient and the extra heat can influence the rate of larval development. If a larval mass is noted, it should be photographed and the mass temperature should be taken, prior to the location being sampled. The temperature of every maggot mass should be taken at each site on the body, so that this can be taken into consideration when calculating the crime scene thermal history.

5.2.2 Pupae and puparia

Fly puparia are usually found some distance from the body. The third instar post-feeding larval stages migrate and can be found in soil 3–5 cm below the soil surface, in pockets, under carpets, in leaf litter or in any nooks and crannies which are available in buildings. If the puparia are still on the body, then either there may have been some restriction to larval migration, or a particular species of insect is indicated. Puparia change colour from white to dark brown over time, so all puparia, of whatever colour, should be recovered.

An organized search strategy should be used to do this. The ideal is to search on a grid of a metre apart over a 36 square metre area surrounding the body, if it is not in a house. This is a slow, time-consuming activity in which the soil should be sampled at the intercepts of the grid, using a trowel to a depth of 10 cm. The soil may need to be sieved over a tray, or it can be hand-searched. As previously indicated, the puparia recovered are placed in a container with a moist paper towel and suitably labelled. They do not require feeding but should be taken back to the laboratory for identification. The puparia should be cultured through to emergence if at all possible, so that species identification can be confirmed. The puparial case should also be retained as additional evidence. Those which do not hatch provide the examples of preserved specimens from the scene.

5.3   Catching adult flying insects at the crime scene

Flying insects present at the scene should be collected first using a net, before hand-collecting any specimens from the body. This is because they are most easily captured using a net and may disappear if disturbed. The net is flicked from behind the insect in an upward sweep, catching it within. Then, with a wrist swing, the net should be folded over at the end to contain the insect. At this point the bag can be grabbed with the other hand (which hand depends on whether you are left- or right-handed) and the insect, in the net base, can be restricted so that a container can be placed over it (Figure 5.5). A firm shake usually keeps the insect in the bottom of the tube for sufficient time to put a lid on top.

These insects can either be kept in individual killing jars, or they can be retained until dead in a single killing jar, as a collection of flying insects from the crime scene. Then they are transferred to individual specimen jars. Since insects are mobile, these insects are representative of the crime scene as a whole. In all cases, accurate labelling and recording is imperative.

If the crime scene is a car, relevant evidence can be obtained by collecting any insects which have been trapped on a radiator grill, bonnet or the windscreen (wind shield) of the vehicle. This may provide details of movement of the body. The temperatures in the car may be important, as inside the vehicle is likely to get quite hot and may affect the speed of the insect development, where flying insects have been able to gain entry and lay eggs.

Figure 5.5   Retrieving a flying insect from the net

5.4   Catching adult crawling insects at the crime scene

Insects such as beetles, which are visible on the surface of the body, can be collected by hand-picking and placed in individual, labelled containers. This is a sensible precaution, since beetles may be carnivores and eat any other specimens, thereby destroying your evidence. In an indoor crime scene, it is useful to check the nooks and crannies of the room for crawling insects, as this provides further information about predators and the conditions in which the body has been found. Leaf litter, or ground cover, in an outdoor scene can also be collected at regular points and the contents sieved or again hand-picked. Pitfall traps can be used to catch crawling insects near the body if it is an outdoor crime scene.

Figure 5.6 A commercial Tulgren funnel for extracting live insects from soil

Tulgren funnels (Figure 5.6) can be used to recover the soil organizms which are living under the body. Several samples of soil (around 5 g each) are collected. Each is placed in a Tulgren funnel and a light positioned above the sample. As the soil dries out, the organizms are driven down into the container of 70% alcohol below. These can later be identified to give a profile of the specimens below the body and elsewhere at the crime scene.

5.5 Obtaining meteorological data at the crime scene

It is extremely important to determine the temperature at which the insects were growing on the body before it was discovered. The estimates of time since death rely on the figures gleaned at the crime scene and those determined subsequently from other sources.

The body temperature should be determined by placing a thermometer on the body surface. The temperature of the air should be taken at a height of 1.1 metres

(4 feet); this provides a measure of ambient air temperature at a comparable height to those taken at the meteorological station. Care should be taken to avoid holding the actual thermometer; use a protector or a rubber band wound round the end. Do not expose the thermometer to direct sunlight, as this will raise the temperature and give a false reading – your body may provide some shade. The temperature directly beneath the body should be taken, followed by the soil temperature. To take the temperature of the soil, it is better to use a soil thermometer so that there is little chance of the thermometer breaking as it is forced into the ground.

A copy of a possible protocol for the collection of specimens from the crime scene is presented in Appendix 1.

5.6 Review technique: investigating the influence of larval location

5.6.1 Introduction

Thigmotaxis, or the desire for physical contact, is a particular feature of dipteran larvae and influences their tendency to both form larval masses and also to 'hug' the sides of containers. The 'maggot mass' can raise the temperature well above ambient. Catts (1992) recorded a temperature of a 'maggot mass' 20°C greater than the highest ambient temperature at the time. This practical task allows you to demonstrate this aspect of larval behaviour and to relate numbers of larvae to temperature (a video clip of maggots demonstrating thigmotaxis is on the website at: www.lincoln.ac.uk/entomology).

The experiment below also simulates the effect of different numbers of larvae at different colonization sites on a body at a crime scene, so that you are aware of why this has to be taken into consideration when deciding the ambient temperature experienced by the larvae.

5.6.2 Safety instructions (COSHH)

- Fluon can be an irritant if it touches the eyes or skin.

- Wear gloves and a laboratory coat for this activity and observe normal laboratory etiquette.

5.6.3 Materials

15 Boiling tubes
Alcohol thermometer

Pen or indelible marker
1800 late second instar larvae
Labels
Fluon, 50:50 mixture with distilled or deionized water

5.6.4 Method

1. Take five boiling tubes and place a layer of diluted Fluon around the top of each, to prevent the larvae escaping.

2. Into the first tube, place 40 larvae; into the second, 80 larvae; into the third, 160 larvae; and into the fourth, 320 larvae.

3. Keep the final tube empty. This is your control to determine the ambient temperature.

4. Repeat this distribution of larvae into two lots of five further sets of boiling tubes, so you have three replicates of each number of larvae.

5. Place the boiling tubes away from direct sunlight, so that you have randomly assigned numbers of larvae side by side, in different rows. This ensures that you do not have any bias in your results because of position of the boiling tube.

6. Record the temperatures of each mass of maggots by inserting a thermometer into the mass after 30 minutes of equilibration.

7. Record the ambient air temperature by taking the temperature of the fifth tube at the same time as you sample the larval temperatures.

8. Repeat this temperature assessment after a further period of 30 minutes.

9. Calculate the mean and standard deviation of the temperatures for the populations of maggots for the three replicates of 30 minutes.

10. Present the results of your experiment as a graph of larval numbers (x axis) against temperature (y axis). What did you note about the effect on temperature of the increased larval numbers?

5.7   Further reading

Adams Z. J. O. and Hall M. J. R. 2003. Methods of the killing and preservation of blowfly larvae, and their effect on post mortem larval length. *Forensic Science International* **138**: 50–61.

Byrd J. H. and Castner J. L. 2001. *Forensic Entomology: The Utility of Arthropods in Legal Investigations*. CRC Press: Boca Raton, FL.

Catts E. P. and Haskell N. H. (eds). 1990. *Entomology and Death: A Procedural Guide*. Joyce's Print Shop: Clemson, SC.

Clark K., Evans L. and Wall R. 2006. Growth rates of the blowfly, *Lucilia sericata*, on different body tissues. *Forensic Science International* **156**: 145–149.

Dadour I. and Cook D. 2005. *Insects and Forensic Entomology: Flies Commonly Associated with Corpses in Western Australia*. University of Western Australia: Perth.

Lord W. D. and Burger J. F. 1983. Collection and preservation of forensically important entomological materials. *Journal of. Forensic Sciences* **28**(4): 936–944.

Stubbs A. and Chandler P. 1978. *A Dipterist's Handbook*. The Amateur Entomologist 15. The Amateur Entomologists' Society: London.

Tantawi T. I. and Greenberg B. 1993. The effects of killing and preserving solutions on estimates of larval age in forensic cases. *Journal of Forensic Sciences* **38**: 303–309.

6

Breeding entomological specimens from the crime scene

Once collected, the larvae of flies and beetles should be reared to the adult (imago) stage in the laboratory. This should be undertaken by the forensic entomologist so that the identity of the species of insect can be confirmed. Rearing from the egg stage onwards, carried out under conditions that simulate the crime scene, allows an accurate post mortem interval to be estimated.

As each life stage is reached in the laboratory, samples developing from those recovered from each location on the body (if possible at least 20 per sampling) should be despatched in boiling water and preserved in Kahle's solution. Data relating to details of temperature and time to reach this stage should also be recorded, both in a laboratory note book and also on the sample pots, so that the pots collected at the crime scene can be related to this information. This record, along with the specimens, may be requested by the court, or be used in court to illustrate your methodology.

Insects collected for culturing from a crime scene must be kept in conditions which enable them to grow successfully. Whilst all of the insects feed on carcasses, it is not appropriate, from both the health and safety standpoints or in terms of human tissue retention laws, to utilize flesh from the corpse to feed these specimens in the laboratory. A supply of food, e.g. pig's liver, will have been brought to the crime scene to provide food for the samples of living larvae collected from each site on the body. This should also be the food used for the rest of the culturing period.

6.1 Returning to the laboratory with the entomological evidence

The conditions under which the insects are kept are extremely important. Specimens must always be kept at the laboratory in conditions which allow them to be stored without damage, or to be grown through their life stages. The larvae from each of the locations on the body should be separated from each other, in separate containers.

The possible influence of the temperatures when the specimens are being transported to the laboratory should always be taken into consideration. If the samples were collected by a crime scene investigator, they should have been brought to the laboratory from the crime scene at a temperature below the base temperature for those insects likely to be found at the crime scene and passed to the forensic entomologist as rapidly as possible. This temperature may need to be determined for local conditions. Myskowiak and Doums (2002) point out that temperatures as low as the normal refrigeration temperature of 4°C may cause alteration of the duration of life stages and the time taken to reach the adult stage. They specified that 10 days of refrigeration prior to the culturing conditions led to an alteration of 9–56 hours from the normal 15.5 days it took when the larvae of *Protophormia terraenovae* were reared at 24°C. Therefore, transit temperatures should be taken into consideration, particularly if there is any deviation from the expected life cycle in the laboratory.

6.2   Fly-rearing conditions in the laboratory

Where larvae are recovered from the crime scene, the foil packages of meat, each containing around 50 larvae, are placed in containers such as polystyrene cups. These can be stored in the dark in the controlled environment cabinet or room (Figure 6.1), until the larvae reach the post-feeding stage. A pierced lid is placed on the top of each container to reduce drying out; one per sample from

Figure 6.1 Larvae collected from the crime scene are reared in a controlled environment cabinet. They are kept in pots

each location, for each type of larvae collected. The pierced lid allows some gas exchange and prevents the build-up of ammonia as the larvae grow. These pots should be maintained at a relative humidity of 65 % or in baths of water at the appropriate temperature, so that the microclimate around the containers prevents the eggs, or initial larval instars, from drying out. Work by Introna *et al.* (1989) confirmed that being reared in growth cabinets under conditions reminiscent of the wild does not statistically alter the life cycle durations of flies. In this instance they used *Lucilia sericata* for the trials.

As a precaution against loss and intermixing of cultures, each pot can be placed in a second container, such as an aquarium. The sides of the aquarium should either be treated with a layer of Fluon (50:50 with water), so that the larvae cannot gain a purchase, or covered with mesh. A valuable alternative to mesh over a cage is a pair of tights with the feet cut off. These cover the container and prevent any of the maggots escaping from the containers, whilst allowing a good air flow. Each container should be clearly labelled with date, case, collector and item number.

Adult flies should be maintained in large cages (Figure 6.2) in order to mate, to gain eggs for development through to the stage recovered at the crime scene. These cages

Figure 6.2 Large cage for emerging adult flies

should be around $46 \times 36 \times 46$ cm and have a mesh cover to allow light to enter and air to circulate, whilst retaining the culture. If the cage is too small, the insects' wings will become battered and flight will be affected, so mating will not be able to take place successfully. Access to the cage, so that food can be replaced, is through a 'sleeve' in the front of the cage. The adults are provided with a constant supply of water, either as a screw-top jar with a wick emanating through the top, or in a Petri-dish with water and stones, or on a sponge so the flies can drink and not drown.

The nutrient supply for the adults should be a 50:50 mix of sugar and water if the culture is to be retained (if sugar alone is used, the insects may potentially become reduced in size, although artificial diets have been successfully used to rear calliphorid larvae). Meat, or liver (Figure 6.3) is also placed in the cage, both to provide nutrients for ovary development in the females and also as a place where the eggs are laid. This meat can be minced (ground) or can be palm-sized pieces of liver. Porcine liver has been used most successfully; ox or sheep liver can also be used, but care should be taken to be consistent in the type and source of meat used. Work by Kaneshrajah and Turner (2004) showed a reduction in the rate of larval growth of *Calliphora vicina* on liver, in contrast to growth when heart, lungs, kidney or brain tissue were used. However, the majority of research workers successfully use *ad libitum* liver as the food source for carrion feeders from forensic sites, without any effect on the duration of life cycle stages. Hermes (1928), however, showed that an effect on the sex ratio of *Lucilia sericata* was caused by the amount of food available to the larvae, so *ad libitum* feeding is

Figure 6.3 Container with liver provided as an oviposition site so that the larvae can be grown through to the stage recovered from the body at the crime scene

Table 6.1 Average minimum life cycle duration (hours) of a selection of dipteran species at fixed temperatures

Species	Temperature (°C)	Egg stage	L1	L2	L3*	Pupariation	Source
Calliphora vomitoria (L.)	12.5	64.8	55.2	60.0	434.4	717.6	Greenberg and Kunich, 2002
	23.0	21.6	25.2	19.2	214.8	247.2	Greenberg and Kunich, 2002; Kamal, 1958
#	26.7	26.0	24.0	48.0	420.0	348.0	
Calliphora vicina (Robineau-Desvoidy)	15.8	41.4	41.6	45.0	166.0	425.7	Anderson, 2000;
	20.6	22.5	34.5	27.0	129.0	301.0	
#	26.7	24.0	24.0	20.0	176.0	288.0	Kamal, 1958
Lucilia sericata (Meigen)	17.0	28.0	39.0	54.0	279.0	442.0	Grassberger and Reiter, 2001
	20.0	22.0	24.0	35.0	161.0	209.0	
Lucilia illustris (Meigen)	21.8	19.3	28.7	45.0	136.0	229.0	Anderson, 2000
	15.0	70.3	75.0	135.0	573.0	458.0	Byrd and Allen, 2001
Phormia regina (Meigen)#	20.0	21.2	30.0	55.0	274.0	244.0	Byrd and Allen, 2001
	25.0	18.9	25.0	44.0	251.0	209.0	Byrd and Allen, 2001
Protophormia terraenovae (Robineau-Desvoidy)	12.5	91.2	290.4	240.0	832.8	722.4	Greenberg and Kunich, 2002
	23.0	16.8	26.4	27.6	118.8	144.0	Greenberg and Kunich, 2002
Sarcophaga haemorrhoidalis (Fallen)	25.0	N/A##	12.0	32.0	112.0	300.0	Byrd and Butler, 1998**;
Sarcophaga bullata (Park)#	26.7	N/A	26.0	18.0	166	288.0	Kamal, 1958

*L3 combines both feeding and non-feeding stages.
**Byrd and Butler (1998) noted a cohort variance that was large and advised the use of broad confidence intervals for *Sarcophaga haemorrhoidalis*.
#Based on mean time (hours) rather than average minimum times in each stage.
##N/A, not applicable – Sarcophagidae are viviparous.

important. The food should be placed in a foil case, or in a tub with a partial lid, so that the area of access at the top can be reduced. This restriction encourages flies to lay eggs and the food to retain its moisture for a longer period. Care should be taken to keep the relative humidity above 50 % and ideally 65 %, in order to stop the eggs drying out.

Flies should be bred at the most appropriate temperature, either in relation to the crime scene or to achieve rapid development. Often this requires a controlled environment cabinet, although a room with a temperature with a limited and recorded fluctuation can also be used. Experimental research provides an indication of a suitable temperature. Information about the expected duration of the life stages comes from a number of sources (Table 6.1), including work by Kamal in the 1950s, who investigated the life cycles of 13 fly species at 26.7°C and 50 % relative humidity (Kamal, 1958), along with papers by Anderson (2000), Byrd and Castner (2001), Higley and Haskell (2001), Greenberg and Kunich (2002) or Donovan *et al.* (2006). The day length which has been successfully used to avoid influencing the life cycle is 16 hours daylight and 8 hours dark (16L:8D; Vaz Nunes and Saunders, 1989). However, the most appropriate day length to use is the average day length for the season in which the specimens were recovered from the crime scene, so that the conditions in the environment prior to recovery of the body are mirrored.

6.3   Conditions for successful rearing to the adult (imago) fly stage

Once the larvae have reached the third instar, they need to be transferred into conditions which ensure that the post-feeding larvae can migrate successfully, whilst preventing loss of the evidence. The ideal is an aquarium with vermiculite, sand or sawdust in the bottom, kept in a controlled environment cabinet at the same temperature as that found at the crime scene. This provides a medium in which larvae are able to bury themselves to pupariate. It is therefore worth containing third instar larvae in an aquarium tank at this stage of their life cycle, by moving the cup, if the aquarium has not previously been used as a means of secondary containment. The extra space is particularly important, since lack of pupariation sites can influence the degree of success in completing the life cycle. As Byrd and Castner (2001) point out, if natural pupariation is prevented, then there may be an extension in the length of time the larvae spend in a particular stage and the PMI estimation will be inaccurate as a result.

6.4   Beetle rearing in the laboratory

For the purposes of rearing, beetles such as Silphidae, Cleridae, Histeridae and Staphylinidae may be kept in transparent plastic containers, glass jars, vials, pots or buckets. They may need to be kept individually to prevent one beetle eating

Figure 1.1

Figure 1.2

Figure 1.3

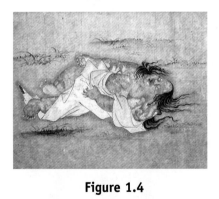

Figure 1.4

Figure 1.5

Figure 1.6

Figure 1.7

Figure 1.8

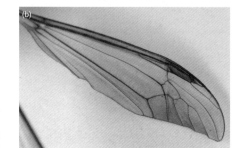

Figure 2.5

Figure 1.9

Figure 2.10

Figure 2.12

Figure 2.13

Stem vein is
bare in Lucilia
species

Figure 2.14

(a)

(b)

Tubercle

Posterior
spiracle

Spiracle
button

Outer rim of
the spiracle is
called the peritreme

The lines are the
spiracle slits

Figure 2.15

(a)

(b)

Figure 2.16

Figure 2.17

Figure 2.18

Figure 2.19

Figure 2.20

Figure 2.22

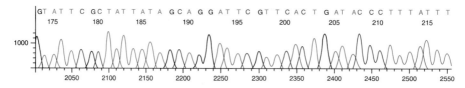

Figure 2.24

Figure 3.3

Figure 3.6

Figure 3.8

Figure 3.11

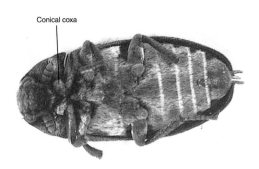

Conical coxa

Figure 3.12

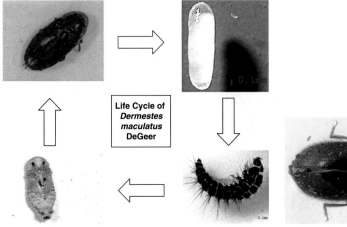

Life Cycle of *Dermestes maculatus* DeGeer

Figure 3.13

Figure 3.14

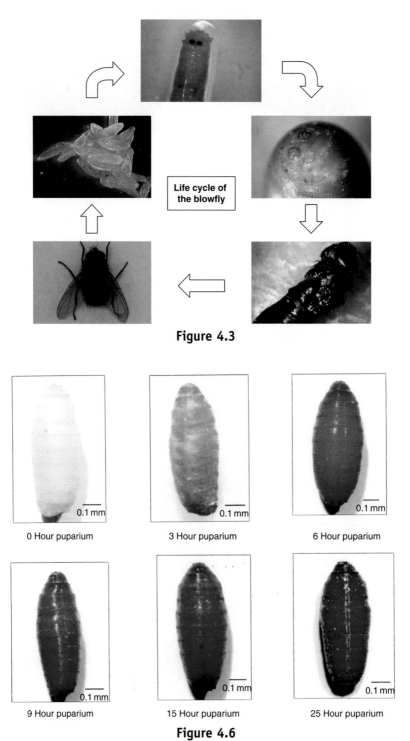

Figure 4.3

0 Hour puparium

3 Hour puparium

6 Hour puparium

9 Hour puparium

15 Hour puparium

25 Hour puparium

Figure 4.6

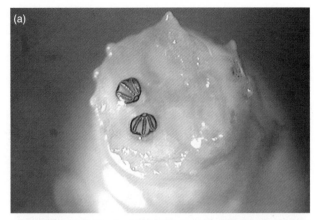

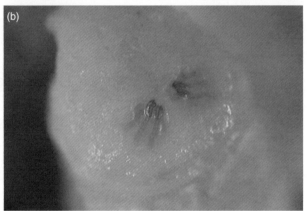

Figure 4.4

Figure 4.7

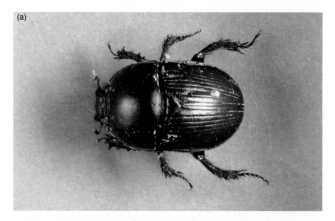

Figure 6.4

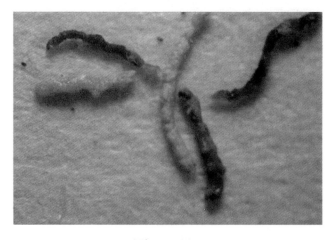

Figure 9.2

another, unless you are setting up breeding conditions, when pairs of beetles should be placed in the containers. These containers should have a layer of moist peat, peat substitute, soil or sawdust in the bottom, depending upon the family of beetles, and places in which insects can hide, such as half plant pots or regions of crumpled paper as a refugium.

Silphid beetles, such as *Microspores* spp. adults, need a temperature of 20°C and a daylight regime of 16:8 (L:D) (Eggert *et al.*, 1998). They should be kept either individually or in a maximum group size of six beetles of the same sex. If the beetles are required to breed, pairs can be placed in a container with a small carcass, such as a defrosted mouse. Alternatively, a large piece of beef, pork or chicken provides enough food (Byrd and Castner, 2001). If you are keeping *Nicrophorus* beetles, the meat left in a container placed on the top of peat, or peat substitute, will be buried and eggs will be laid in the soil near the meat ball (Kramer Wilson, 1999). Such eggs can be recovered and retained on moist filter paper at 20°C until they hatch. Eggs of silphid species such as *Nicrophorus vespilloides* take on average 56 hours to hatch at 20°C (Muller and Eggert, 1990; in Eggert *et al.*, 1998).

Water, in the form of either a Petri-dish with damp cotton wool, or a vial containing water with a wick of paper or cotton wool pushed through the lid, should be provided in the tank for both adult beetles and larvae. The vial is sunk into the peat so that the beetles can readily gain access. The larval culture, each in individual containers, can be supplied with a carcass with a hole in it, by which the larva gains entry. The cultures can be maintained under a regime of total darkness in an incubator or controlled environment cabinet.

6.4.1   Rearing colonies of Dermestidae

Other beetles, such as dermestids, do not require a whole carcass to successfully complete their life cycle and can be bred satisfactorily on dried meat or on an artificial diet. Dermestids breed optimally at a slightly higher temperature than 20°C and are ideally kept at 25°C and 80 % relative humidity (Coombs, 1978). They too can be kept as cultures in aquaria or glass jars. Dermestids require wood sawdust, or some solid medium such as polystyrene or cork, in which to burrow to pupate. This should be covered with several layers of paper to simulate conditions in the body. A supply of water must be provided, either as a piece of folded, dampened paper, or as a container of water with a wick. Keep the water away from the food to avoid encouraging fungal growth. Black paper, in the form of a 'concertina', provides an egg-laying site from which it is relatively simple to see and recover eggs.

6.5   Dietary requirements of insects reared in the laboratory

Both flies and beetles have specific nutrient requirements in order to satisfactorily complete their life cycles. Flies require carbohydrate as an energy source, water and protein. Protein is particularly important to females for the development of the ovarioles and for egg production. They also require a number of vitamins and minerals.

Carrion beetles require a diet which incorporates the relevant nutrients, or to feed on flesh. Therefore, they should be fed dead mealworms every 3–4 days (Eggert *et al.*, 1998). This is a suitable food for ground beetles (carabids), which can also be fed on ant eggs, such as those used for feeding fish, or maggots and pupae. This inclusion of insects in the diet is particularly appropriate for the Silphidae; Cleridae, Histeridae and Nitidulidae can be fed in the same way.

In contrast, larvae and adults of skin beetles (Dermestidae) – *Dermestes maculatus* or *Dermestes lardarius* – can be fed on dried dog food or fish food. They prefer pelleted food rather than flakes, but will consume both. Such food should be checked every 2 or 3 days to ensure sufficient is available and of appropriate quality. The food should be available in excess so that the dermestid life cycle is not influenced.

In order to ensure that, where necessary, dermestids can breed efficiently, it may be necessary to provide meat on an intermittent basis, in order to provide all of the required nutrients for reproduction. To simulate the conditions at the crime scene, using meat such as pork which has dried out may be a suitable food source in preference to using artificial foods.

6.6   Review technique: preserving and mounting insect specimens

This activity provides you with the skills that could be required for presentation of your entomological evidence in court. You will mount both a fly and a beetle adult to provide evidence of the identification features for the species.

Mounting insects (Figure 6.4) is valuable, as it provides examples of specimens of different species of insect, which can be used for illustrative purposes in court. It also provides a voucher collection of specimens for a particular geographic location. This will allow you to spot whether there are any unusual species present and hence the possible movement of a cadaver.

The adults at the crime scene will have been killed by placing them in a killing jar containing ethyl acetate. A new jar should be used for each location on the corpse at which the insects were collected. Those from the general crime scene may all be added to the one jar (ethyl acetate does not cause the insects to become rigid when they die; this means they are easy to position for pinning).

(a)

(b)

Figure 6.4 Examples of styles of mounting insects. (a) A beetle. (b) A fly mounted on card. Small insects such as flies can also be mounted on a point (A colour reproduction of this figure can be found in the colour section towards the centre of the book)

1. Each large insect, e.g. a blowfly, is removed from the killing jar using fine forceps or a fine paint brush. This is so that it can be positioned on a cork or polystyrene tile, or insect mounting block. If using the mounting block, choose one large enough in which to sit the thorax, so that the insect's wings and legs can be spread out.

2. Using a pair of entomological forceps, a pin should be carefully forced through the insect exoskeleton. The pin size should be a size No. 1 and ideally of a non-rusting type.

   • For beetles, the pin is placed through the right elytron, as near as possible to the top margin of the wing case.

   • Fly specimens that are large should have the pin placed in the middle of the thorax (mesothorax region), so that it is in between the wings. Smaller flies may be glued to a card.

3. The legs of both flies and beetles should be positioned so that they are spread away from the body. For example, the first pair of legs of a beetle should be positioned forward, the middle pair of legs should be positioned at 90° to the body and the third pair of legs should be positioned away from the body at a 45° angle to the body. They may need to be held in position on the cork or polystyrene, with their wings and legs retained by small strips of paper with their ends pinned down. Alternatively the fly, or beetle, can be positioned as naturally as possible so that it is the femur which is held in this position and the legs hang down at an angle of 90°. The specimens are left to dry for several days until they are firmly in position. Smaller flies and beetles are less easy to position. They are mounted on a triangular point, using glue. The triangle is usually of 1 cm length and made of white card. Clear nail varnish or Solvate wallpaper paste can both provide a cheap, easily accessible source of glue. The side of the fly or beetle is glued onto the point of the triangle.

The procedure for pinning the specimen and positioning the labels made of white card is the same. This is irrespective of whether the insect is mounted onto a pin directly, or the triangle plus insect is mounted at the mid-point along its base. To explain how to do this, the pinning and labelling of a fly is described:

4. A pinning block is used to position the fly about 1 cm from the head of the pin. Below this, two labels of an appropriate size for the information, and the fly, are positioned in parallel.

5. First, the pin point is pushed into the fly so that it pierces the thorax. The fly is then moved up the pin into position, by locating the pin point into the hole in the highest step of the pinning block and pushing.

6. Next, the fly is removed from the hole in the top step and the first label is placed on the second highest step. The fly plus pin is pushed into the centre of the end of the label, so that the information can be clearly read. The label contains the following information:

   • Name of the collector.

   • Place (crime scene) collected, including where on the body.

- Item number and case number.

- Date collected.

7. The label is pierced by pushing the pin through the label, using entomological forceps and resistance from the block. Once the pin has emerged through the card label, it is slid up the pin to the correct position. This is achieved by pushing the pin down through second hole on the pinning block. It is in position when the pin hits the bottom of the block.

8. A second label indicates the species name of the insect, when it is identified, and who did the identification. This label is put in position using the same sequence of movements, but using the third hole in the lowest step of the pinning block. Both labels are held parallel to each other and are written in indelible ink, or are printed using a laser printer. By using this sequence for mounting insects, the chance of the information about the case number and the location label being lost is lessened. The insect can, after all, be reidentified in the unlikely case of loss of the second label on the pinned specimen.

6.7 Further reading

Byrd J. H. and Castner J. L. 2001. *Forensic Entomology: The Utility of Arthropods in Legal Investigations.* CRC Press: Boca Raton, FL; pp 121–142.

Catts E. P. and Haskell N. H. 1990. *Entomology and Death – A Procedural Guide.* Joyce's Print Shop: Clemenson, SC.

Coombs C. W. 1978. The effect of temperature and humidity upon the Dermestidae. *Journal of Stored Products Research* **14**: 111–119.

Cooter J. and Barclay M. V. L. 2006. *A Coleopterist's Handbook*, 4th edn. Amateur Entomologists' Society: Orpington, Kent.

Kaczorowska E. 2002. Collecting and rearing necrophagous insects, important in determining date of death, based on entomological methods. *Archiwum Medycyny Sajdowej I Kryminologie* **52**(4): 343–350 [in Polish].

Lambkin T. A. and Khatoon N. 1990. Culture methods for *Necrobia rufipes* (De Geer) and *Dermestes maculatus* (De Geer) (Coleoptera: Cleridae and Dermestidae). *Journal of Stored Products Research* **26**(1): 59–60.

Sherman R. A. and My-Tien Tran J. 1995. A simple sterile food source for rearing the larvae of *Lucilia sericata* (Diptera: Calliphoridae). *Medical and Veterinary Entomology* **9**: 393–398.

Sherman R. A. and Wyle F. A. 1996. Low-cost, low-maintenance rearing of maggots in hospitals, clinics and schools. *American Journal of Tropical Medicine and Hygiene* **54**(1): 38–41.

Shin-Ichiro T. and Numata H. 2001. An artificial diet for blow fly larvae, *Lucilia sericata* (Meigen) (Diptera: Calliphoridae). *Applied Entomology and Zoology* **36**(4): 521–523.

7

Calculating the post mortem interval

After identifying the specimens from the body, the next stage is to link this information to the temperature at the crime scene. Temperature data, covering the period since the person was last seen alive, are obtained from the local meteorological station. These data are 'corrected', using a *correction factor* calculated from the meteorological office data and half-hourly temperature readings, which have been recorded at the crime scene for 3–5 days after the body was discovered. These corrected data provide an estimate of the temperatures at the crime scene before the corpse was found. From this information, you can determine the length of time the flies took to grow from an egg to the developmental stage recovered from the body. By implication, this is the best estimate of the post mortem interval (PMI) that is available.

Box 7.1   Hint

Additional comments on PMI

Insects are cold-blooded (poikilothermic) and cannot control their body temperatures, so they use the environment as a source of warmth. Insects use a proportion of the environmental energy (thermal units) to grow and develop. The overall energy budget to achieve life stages can be specifically calculated and its calculation is a common feature of integrated pest management predictions, as well as crop production. The thermal units are called degree days (°D) and can be added together to reflect periods of development. In this case they are called accumulated degree days (ADD). If the period is shorter and the length of time being discussed is in hours then the thermal values will be as accumulated degree hours (°H; ADH).

The minimum temperature for growth (basal temperature) will vary with the species. The maximum temperature is in the region of 52.7°C (126.9°F). The temperature during each 24 hour period may vary but the area under a curve between the upper and lower thresholds of growth represents a predictable block of time-accumulated degree days.

Such estimations of time since death are based on the speed of insect growth. Insects are 'cold-blooded', so their growth is influenced by temperature – below a temperature threshold development stops; above a specific temperature threshold, the rate of growth also slows down. Between these two points, however, the rate of growth of the juvenile insect is considered to have a linear relationship with temperature (Figure 7.1).

The maximum temperature threshold for different species of insect varies. Wigglesworth (1967), for example, suggested the maximum temperature for growth and development for *Calliphora* spp. larvae was 39°C, whilst for *Phormia* spp. it was 45°C. Upper threshold temperatures are rarely experienced when investigating most crime scenes, so this factor is only infrequently important, although if temperatures do remain at, or near, the maximum for a long period of time, this will affect the accuracy of the PMI estimate, as the growth of the insect will be slower than expected. Equally, at particularly low temperatures, development may not be possible at all.

We call the temperature threshold below which growth and development will not take place the *base temperature*. This will vary from species to species and can vary with geographic location. For example, Davies and Ratcliffe (1994) demonstrate a threshold of 3.5°C for *Calliphora vicina* in the north of England, whereas Marchenko (2001), working in Russia, records a base temperature of 2°C for the same species. Donovan, Hall and Turner (2006) explored *Calliphora vicina* growth in London, at temperatures between 4°C and 30°C, and found the base temperature there to be 1°C.

Oliveira-Costa and de Mello-Patiu (2004) point out that calculations which use an inappropriate base (lower threshold) temperature will overestimate the accumulated physiological energy budget – termed *accumulated degree hours* (ADH) or *accumulated degree days* (ADD) – and so the forensic entomologist may give a false post mortem interval. Such species adaptation has to be taken

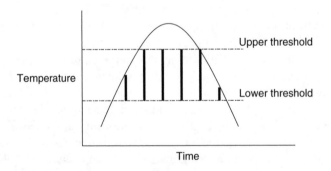

Figure 7.1 Insect growth in relation to upper and lower temperature thresholds. Area outlined is the area of growth. The lower threshold indicates the temperature at which growth is no longer possible. As the temperatures reach the upper limit, the growth rate again slows down. Between the two thresholds insect growth rate is considered linear and hence there is a relationship between temperature and growth rate – or a measure of the growth energy budget requirement

into consideration at crime scenes and the base temperature for common dipteran species in that locality may have to be predetermined.

7.1 Working out the base temperature

The specific base temperature for a particular species is worked out in the laboratory from the insect's growth rate at set experimental temperatures. The calculation is based on the premise that the cooler the temperature, the more slowly the insect develops. The base temperature is calculated by plotting temperature against 1 ÷ total days to develop, i.e. the time between the larvae initially emerging from the egg and the emergence of the adult, using a range of temperatures. If the line of the graph is extended down to the x axis, the point where it meets the x axis (abscissa) can be read off (Figure 7.2). This is the base temperature for that particular species. This graphical method of determining the base temperature is called the *linear approximation estimation* method.

Box 7.2   Hint

How to get the temperature correction factor

- Enter your columns of temperature data from the crime scene and meteorological station of the 3–5 days in the EXCEL work book.

- Click on CHART WIZARD on the tool bar.

- Click on XY Scatter-style graph.

- Press NEXT.

- Enter the titles for meteorological station (°C) in Value (x) axis and crime scene (°C) in Value (y) axis and press NEXT

- Choose whether you want the data as a separate chart or to be on the sheet with your data, then press FINISH

- RIGHT CLICK on one of the points on the graph and choose ADD TRENDLINE

- Choose options DISPLAY EQUATION and DISPLAY R^2 VALUE

- Locate the equation so that it is easy to read in relation to the graph. This will give you the correction factor for the crime scene temperatures before the body was found.

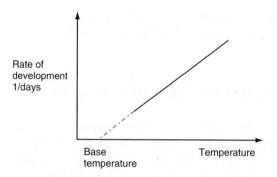

Figure 7.2 Base temperature determination, using the linear approximation method. The point at which the graph extends to the x axis is the approximation of the base temperature

7.2 Accumulated degree data

As we have seen from Figure 7.1, there is a relationship between rate of insect growth from the egg stage to adulthood and temperature. This is because growth and development through the various life stages has a cost in terms of a 'physiological development energy budget'. This budget can be expressed in thermal units called *degree days* (°D) or *degree hours* (°H).

The methods for working out degree days, or degree hours, range from using averages through to using such transformations of the temperature as sine wave, cosine wave and integration calculations. Work on accumulated degree days at the University of California (Wilson and Barnett, 1983) suggests that these methods are interchangeable and for most calculations there is little variation in level of accuracy between those which use transformations and those which are based on average figures. Therefore, for the sake of simplicity, the averaging method for a linear estimation of ADD or ADH will be described here, as this can be applied to both indoor and outdoor crime scenes.

The hypothesis on which insect growth in degree days is based is that, between the upper and lower thresholds, the rate of growth of the insect is linear in relation to temperature increase. This 'physiological energy budget' can be represented as the area under a curve, for temperatures above the base temperature, in each 24 hour period. As can be seen from the graph (Figure 7.3), for each hour or day, the budget is represented as a rectangle of time in relation to temperature; any underestimate at one point in the accumulation is compensated for by an overestimate at another point on the graph. Therefore, total accumulated degree hours (or days), reflects the time taken for the insect to develop to the stage recovered from the crime scene. Based on this relationship, accumulated degree hours (ADH) (or days, ADD) can be determined from a formula. The formulae are:

$$\text{Time}_{(\text{hours})} \times (\text{temperature} - \text{base temperature}) = \text{ADH}$$

$$\text{Time}_{(\text{days})} \times (\text{temperature} - \text{base temperature}) = \text{ADD}$$

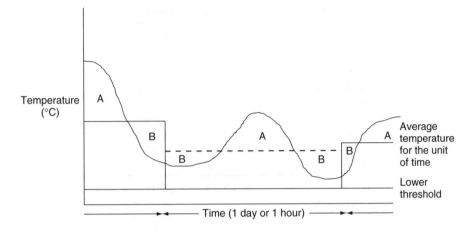

Figure 7.3 Graph to show the justification for using accumulation of averaged temperatures over time. The method assumes a straight line relationship between the chosen periods of time, on the basis of averaging the temperature over the time period. The assumption is that overestimates at one point (A) are compensated for by underestimates at another point (B) within that same time period. Hence, averages for each hour (or day), provided they are based on a number of readings, should be accurate. Modified with permission from Wilson and Barnett (1983),℠ © 1983, Regents, University of California

Information about time to complete individual stages, at set experimental temperatures, comes from the literature. The sources range from Kamal (1958), Vinogradova and Marchenko (1984), Byrd and Butler (1998), Anderson (2000), Greenberg and Kunich (2002) to Lefebvre and Pasquerault (2004). These experimental temperatures are multiplied by the time, usually in hours, taken to reach the individual life stages. For example, the duration of the egg stage cited in the literature, plus first instar duration, plus second instar duration, and so on, are all added up to provide a total experimental time period to reach a particular stage in the life cycle (the number of life stages which have to be taken into consideration is predetermined by which stage was found on the body). The base temperature (Table 7.1 provides examples) must be subtracted from the temperature at which the specimens were grown, before multiplying this figure by the time taken to pass from the egg stage to the chosen life cycle stage.

Once we have identified the species and worked out the experimental energy budget to reach the life cycle stage recovered from the body, we need to turn to the conditions at the crime scene. The physiological energy budget (accumulated degree hours or days, i.e. ADH or ADD), which was built up over time at the crime scene, has to be worked out for the period between death and discovery of the body. It is based on the individual temperature fluctuations at the crime scene within each 24 hour period, as either an average per hour or a daily average. Had the temperatures at the crime scene prior to the discovery of the body been known, these thermal units could merely be added up until the point where the summation

Table 7.1 Lower development thresholds

Species	Base temperature (°C)
Calliphora vicina	2.0
Calliphora vomitoria	3.0
Protophormia terraenovae	7.8
Lucilia sericata	9.0
Chrysomya albiceps	10.2
Phormia regina	11.4
Muscina stabulans	7.2

Reproduced from Marchenko (2001) with permission from Elsevier, © 2001.

approximated to the experimental 'physiological energy budget' (ADH or ADD), for that species. However, we rarely have these data for the period before the body is discovered, so crime scene temperatures have to be estimated from the information which is available. Usually these are the data from the nearest meteorological station. Each daily or hourly energy budget is calculated by multiplying by one, the temperature from which the base temperature has been subtracted.

It is also most important, when calculating the ADH or ADD from the scene, that the experimental temperatures used are in the same units as those used for recording the crime scene temperatures. If degrees Fahrenheit (°F) were used, then this should also be used for recording the temperature at the crime scene. If degrees Centigrade (°C) were used, then this should be the measure used. The units in which the temperature is recorded at the local meteorological station usually dictates what is used, but the figures may, on rare occasions, need to be converted.

Ideally, the temperature measurements for the time since the victim was last seen should be based on hourly averages, and you should calculate an ADH (°H) measure of post mortem interval. However, in practice this may not be possible because the available meteorological data is given only as daily maximum and minimum temperatures. Under such circumstances, you should calculate accumulated degree day measurements (ADD or °D), because they reflect the level of accuracy of the PMI estimate that can be ascribed. (Remember, you can convert ADH to ADD by dividing ADH by 24. You cannot accurately convert from ADD to ADH.)

The choice of whether to use ADD or ADH is also dictated by the level of accuracy which is most appropriate. As a rule of thumb, calculations should be in accumulated degree hours (ADH) if the victim has been missing for less than a month. If they have been missing for more than a month, the data is more appropriately presented as accumulated degree days (ADD). This is because the accumulated level of variation in shorter time periods gives a less accurate measure of the 'physiological energy budget'.

Box 7.3 Hint

Converting temperatures

In order to convert a temperature in Fahrenheit (T_f) to Celsius (Centigrade) (T_c), subtrsact 32 from the temperature in Fahrenheit, multiply the result by 5 and divide by 9:

$$T_c = (5 \div 9) \times (T_f - 32)$$

In order to convert a temperature in Celsius (Centigrade) to Fahrenheit, multiply the temperatures in Celsius by 9, divide the answer by 5 and then add 32 to the total:

$$T_f = [(9 \div 5) \times T_c] + 32$$

7.3 Calculation of accumulated degree hours (or days) from crime scene data

7.3.1 How to obtain corrected crime scene temperatures

Corrected data for the crime scene before the body was found come from comparing temperatures from the local meteorological station with those for the crime scene, once the body has been discovered. A scatter diagram is plotted of the meteorological temperatures (x axis) against the crime scene temperatures (y axis) recorded for 3–5 days after the body was discovered. A regression equation is calculated. This equation is then used to correct each of the meteorological station readings to generate predicted crime scene temperatures (Figure 7.4). These corrected average hourly, or daily, temperature readings are used in the calculation of accumulated degree hours (ADH) or accumulated degree days (ADD).

Because the temperatures are based upon hourly or daily averages, the time used is either 1 hour or 1 day. Each of the above figures (minus the base temperature) is multiplied by 1, i.e. 1 hour or 1 day. Then each result is added to the former accumulated figure, working backwards from the time of discovery of the body, until the figure for the experimental ADH or ADD is reached. The number of days or hours to reach this figure is then counted up.

An EXCEL spreadsheet can be used to input the initial Meteorological Office data, work out the best estimate of crime scene temperature for the species of insect, subtract the base temperature and obtain a figure for the post mortem interval. This approach means you calculate the figures quickly and accurately. Table 7.2, together with the Hint on 'The post mortem interval calculation', provide

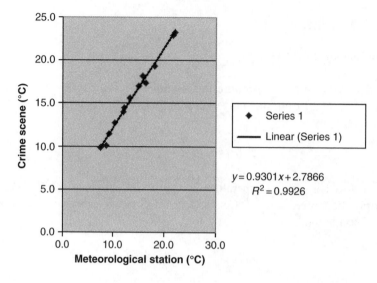

Figure 7.4 Regression of crime scene data against meteorological station temperature data to determine the correction factor for 20–23 April ($y = 0.9x + 2.8$)

Table 7.2 An example of the headings and completed spreadsheet to calculate ADD for *Calliphora vicina* (Robineau-Desvoidy)

Species	Daily meteorological temperature data*	Crime scene temperature data*	Base temperature*	ADD	Σ ADD
Calliphora vicina	7.6	10.4	2.0	8.4	
	9.6	12.4	2.0	10.4	18.8
	10.2	13.0	2.0	11.0	29.8
	12.0	14.8	2.0	12.8	42.6
	12.4	15.2	2.0	13.2	55.8
	12.6	15.4	2.0	13.4	69.2
	13.2	16.0	2.0	14.0	83.2
	12.8	15.6	2.0	13.6	96.8
	14.1	16.9	2.0	14.9	111.7
	13.9	16.7	2.0	14.7	126.4
	15.0	17.8	2.0	15.8	142.2
	15.2	18.0	2.0	16.0	158.2
	13.7	16.5	2.0	14.5	172.7
	13.9	16.7	2.0	14.7	187.4
	14.3	17.1	2.0	15.1	202.5
	12.8	15.6	2.0	13.6	216.1
	12.7	15.5	2.0	13.5	229.6
	13.8	16.6	2.0	14.6	244.2
	14.3	17.1	2.0	15.1	259.3
	15.1	17.9	2.0	15.9	275.2

15.4	18.2	2.0	16.2	291.4
14.8	17.6	2.0	15.6	307.0
14.7	17.5	2.0	15.5	322.5
13.2	16.0	2.0	14.0	336.5
12.8	15.6	2.0	13.6	350.1
13.0	15.8	2.0	13.8	363.9

*Units for temperature measurement (°C or °F) must be consistent throughout the calculation.

Box 7.4 Hint

The post mortem interval calculation

Column	Information	Source
A	Species	Field work
B	Meteorological station data	Field work or meteorological station
C	Crime scene data	Determine using regression from work at scene after the body is discovered
D	Base temperature	From literature based on geography
E	Accumulated degree hours (ADH) or accumulated degree days (ADD)	C2-base temperature calculated by EXCEL
F	Sum of ADD or ADH	Sum (E2:E3) calculated by EXCEL

Information for making PMI tables:
Put the following formula into the initial row of boxes on the EXCEL work book:

Column B – put meteorological station temperatures.
Column C – put = B2, followed by the regression equation suitably modified for EXCEL.
Column D – put the base temperature.
Column E – put the formula = C2 – D2 to provide the ADD or ADH figures.
Column F – in the 2nd box down, put SUM E2:E3, and in the 3rd box down, put the formula = F3 + E4, so that you can read off the accumulated degree totals from column F.

Copy the formula down the columns by pulling on the right-hand box, so that you have the answers for each hour or day of accumulated energy budget.

an example of such a table, complete with figures and instructions on how to make it. The ADD and/or ADH must be calculated for each species of fly present on the body. When considered together, these data provide confirmation of the predicted post mortem interval that you have calculated.

7.4 Sources of error

When calculating the post mortem interval, a number of factors need to be taken into account. It is important to consider using the temperature of the maggot mass as the temperature for larval development in particular instars. If maggot mass temperature was recorded as greater than the ambient temperature, the temperature of the mass should be used in the calculations. This is true where third or potentially late second instar larvae are recovered from the body, as the maggot mass temperature may be the highest temperature experienced by the larvae (Higley and Haskell, 2001). If puparia are recorded, the crime scene soil temperature at 5, 10 and 20 cm depth should be used to adjust the estimated crime scene air temperatures, for the period likely to reflect the time the insect was in pupariation.

Where there is no experimental growth data available for the particular species, the larva should be reared until the adults are mature and oviposit. The eggs can then be maintained at a temperature which represents that estimated for the crime scene. The duration from egg stage to the stage of the life cycle which was recorded at the crime scene, will provide a means of estimating the post mortem interval and also of providing confirmation of any post mortem interval which has been calculated.

Base temperatures must be considered for individual species and the correct base temperature must be used. It may be necessary to use several base temperatures to calculate the post mortem interval, in order to reflect the other information relating to the case, particularly when the person was last seen some distance away or the body may have been moved (if the temperature for the period being considered is below that of the base temperature, then a value of zero is included in the calculation for the particular hour or day).

Concern about the accuracy of temperature predictions has been expressed where the meteorological station temperature recordings, for the period before the body was discovered, are at variance with the records for the few days or weeks after the body was discovered (Archer, 2004). If the weather conditions differ markedly, the temperatures during the 3–5 days when the temperature is recorded at the crime scene may not give an accurate reflection of the crime scene conditions. 'Correcting' these data will not produce a sufficient level of accuracy between the two sites. The best way forward in this instance is to grow the insects through to the life stage recorded from the body, using the crime scene temperature.

Wall (2004) expressed concern about using average daily temperatures when calculating accumulated degree data, rather than taking account of the temperature fluctuations. He pointed out that, in his experiments in 2003, using average (mean) temperatures gave considerably longer estimates for insect development (ADD) than when temperature estimates were based upon maxima and minima to determine ADD.

7.5 Use of larval growth in length to determine post mortem interval (isomegalen and isomorphen diagrams)

Where the corpse has been discovered indoors, or in a controlled environment where the temperature is not fluctuating, the relationship between temperature and growth can be used in another way. Under such conditions, the length of the larva, when killed in the standard way by immersing in boiling water, can be related to the time since the larva hatched. Graphs are produced, under controlled conditions in the laboratory, for the time since hatch of the species against the average minimum length. The time since hatch can then be read directly off the graph on the basis of the length of the individual larvae collected from the crime scene. These graphs are called *isomegalen diagrams* and have been calculated for *Lucilia sericata, Protophormia* (= *Phormia*) *terraenovae* and *Calliphora vicina* (Grassberger and Reiter, 2001, 2002; Reiter, 1984).

A second type of graph can be used, which is derived where life cycle stages from egg hatch to the time of emergence of the adult (eclosion) have been plotted against time, at specific temperatures. Each line indicates a change in the life cycle to the next stage. The areas between the lines relate to the identical morphological stages. These are called *isomorphen diagrams* and they have been calculated for the same three species as have the isomegalen graphs. Isomorphen diagrams are useful when post-feeding larvae and/or puparia are collected from the crime scene. From these stages, the post mortem interval can be read directly off the graph, provided that the temperature has been constant.

7.6 Calculating the post mortem interval using succession

Investigating post mortem interval for a period of at least 3 or more months may mean that there is a large assemblage of flies, beetles and other insects present on the body. These can be used for the calculation of PMI using another method. This method requires that first of all every specimen is identified to family. After that, an attempt is made to relate this 'snap-shot' of decomposition fauna to the succession of insects which routinely colonize a corpse at that site. Knowing which insects are present and which are absent locally, in what season, helps the entomologist to estimate the post mortem interval.

7.6.1 Movement of the corpse

The particular assemblage of insects present on a corpse is also an important indicator of whether or not the body has been moved. If an unexpected species is present, which is more characteristic of a different habitat or geographic region, then the body may have been moved. This again depends on knowledge of the local fauna. Organizations such as local wildlife trusts, nature reserves or amateur naturalist societies can be a source of important information about species which are expected in a particular area. Back copies of their house journals may provide published accounts that have received peer review and can provide a basis for your conclusions which will be acceptable to a court.

7.6.2 Predators feeding on insects infesting a corpse

The longer a body remains undiscovered, the greater the chance that insects such as wasps and ants will consume those insects that are feeding directly on the body. This destruction of evidence can cause an interpretation problem relating to time since death. Ants, for example, may carry away eggs and the population of the next generation of colonizers may be reduced as a result. Equally, beetles such as staphylinids and carabids may feed on the adults and larvae which are present on the body. Sometimes feeding takes place at night, so you will be less aware of their presence; others will feed on the younger life stages or attack adults during daylight. In either case, there will be an alteration in the sequence of the succession of insects, and some species which would be expected to be present, may not appear. This information about predation can be important when interpreting the data, if the individual has been dead for a period greater than a couple of days.

7.7 Review technique: interpretation of data from a crime scene case study

7.7.1 Introductory background for the assignment on a fictitious murder

The body of a young girl was found on 20 April at 11 am in the Pleasure Gardens, in a small seaside resort called Wingsea. There were signs of a struggle, but no cuts or knife wounds were noted on the body by the pathologist who attended the scene. Neighbours had last seen the girl in the previous week. Larvae were recovered from beneath the eyelids and in the nasal cavities of the corpse. Thirty larvae were collected from each of these locations and preserved for future identification. In

Box 7.5 Hint

Calculation of the ADD for a corpse found in an urban location

This data is based on a crime scene with a corpse from which early third instar larvae were collected. Table 7.2 illustrates the approach using an EXCEL sheet to calculate accumulated degree days (ADD), when only the daily average temperatures are available:

$$ADD = time_{(hours/24)} \times (temperature - base\ temperature)$$

Regression equation, calculated from the assessment of the temperatures at the crime scene relative to the temperatures at the meteorological station after the body has been discovered, must be used to provide a correction factor.

In this instance, the temperatures had been collected by the Police and a regression analysis undertaken to relate crime scene temperatures to those at the meteorological station. The regression equation (and hence the correction factor) was calculated to be $x + 2.8$. Therefore 2.8 can be added to each met-station daily average temperature ($x + 2.8$) without modification. The equation from the regression analysis will indicate what the factor is. For this example I have chosen one similar to the correction factor figure which was calculated for the Wingsea case study. It will be different for each location.

The base temperature for *Calliphora vicina* which was chosen as most applicable is 2°C, using figures from Marchenko (2001). The forensic entomologist should either determine his/her own base temperature for the particular geographic location or choose the most appropriate estimate for the species from the literature.

Using the data from Kamal (1958), because at the time the temperatures at the crime scene were in the mid-20s (°C), the following times apply for each of the stages (in hours):

Egg stage   24.0
L1   24.0
L2   20.0

The total experimental time period to reach the third instar is 68 hours.
The ADH is $68.0 \times 24.7 = 1679.6$
The ADD is $1679.6 \div 24 = 70.0$

From Table 7.2, and working down to the nearest value for the Σ ADD, the sixth day previous to the discovery of the body is the most likely estimate for the time of death.

addition, a further 30 larvae, from both under the eyelids and in the nasal cavities, were collected to be cultured in the laboratory.

Meteorological station temperature readings for 20–23 April were obtained. Temperatures for the same time period were recorded at the crime scene by the police, using a portable weather station. A regression equation for the relationship (Figure 7.4) between crime scene temperature and meteorological station temperature, was calculated to be $0.9x + 2.8$. This will be used as the correction factor for meteorological station data (Table 7.3), working backwards from the date the body was discovered.

The species from both beneath the eyelids and in the nose were later identified as second instar larvae of *Calliphora vomitoria* L. The entomologist grew the second instar larvae to the adult stage in the laboratory, in order to confirm this identification. A base temperature for *Calliphora vomitoria* of 3°C was chosen.

7.7.2 Instructions

Interpret the data in Table 7.3, to provide an estimate of time since death of the victim. Use data from Greenberg and Kunich (2002) for the length of the *Calliphora vomitoria* life cycle.

Table 7.3 Temperature readings from the meteorological station for the period 14–20 April

Time	Temperature (°C)	Time	Temperature (°C)	Time	Temperature (°C)
20/4 11 a.m.	12.0	7 a.m.	6.0	5 a.m.	7.0
10 a.m.	11.0	6 a.m.	6.0	4 a.m.	7.0
9 a.m.	9.0	5 a.m.	6.0	3 a.m.	6.0
8 a.m.	8.0	4 a.m.	6.0	2 a.m.	8.0
7 a.m.	6.0	3 a.m.	6.0	1 a.m.	9.0
6 a.m.	4.0	2 a.m.	6.0	15/4 12 Midnight	9.0
5 a.m.	4.0	1 a.m.	6.0	11 p.m.	9.0
4 a.m.	4.0	17/4 12 Midnight	6.0	10 p.m.	8.0
3 a.m.	4.0	11 p.m.	7.0	9 p.m.	9.0
2 a.m.	4.0	10 p.m.	9.0	8 p.m.	11.0
1 a.m.	5.0	9 p.m.	9.0	7 p.m.	12.0
19/4 12 Midnight	5.0	8 p.m.	10.0	6 p.m.	13.0
11 p.m.	6.0	7 p.m.	10.0	5 p.m.	14.0
10 p.m.	6.0	6 p.m.	12.0	4 p.m.	14.0
9 p.m.	6.0	5 p.m.	13.0	3 p.m.	14.0
8 p.m.	7.0	4 p.m.	13.0	2 p.m.	14.0

Time	Temp	Time	Temp	Time	Temp
7 p.m.	10.0	3 p.m.	14.0	1 p.m.	14.0
6 p.m.	12.0	2 p.m.	14.0	12 Midday	13.0
5 p.m.	12.0	1 p.m.	13.0	11 a.m.	13.0
4 p.m.	12.0	12 Midday	13.0	10 a.m.	12.0
3 p.m.	12.0				
2 p.m.	12.0	11 a.m.	12.0	9 a.m.	9.0
1 p.m.	12.0	10 a.m.	11.0	8 a.m.	7.0
12 Midday	11.0			7 a.m.	6.0
11 a.m.	11.0	9 a.m.	8.0	6 a.m.	4.0
10 a.m.	10.0	8 a.m.	7.0	5 a.m.	4.0
9 a.m.	9.0	7 a.m.	7.0	4 a.m.	5.0
8 a.m.	8.0	6 a.m.	7.0	3 a.m.	7.0
7 a.m.	7.0	5 a.m.	7.0	2 a.m.	7.0
6 a.m.	4.0	4 a.m.	7.0	1 a.m.	7.0
5 a.m.	4.0	3 a.m.	7.0	*14/4* 12 Midnight	9.0
4 a.m.	4.0	2 a.m.	8.0	11 p.m.	10.0
3 a.m.	4.0	1 a.m.	8.0	10 p.m.	11.0
2 a.m.	4.0	*16/4* 12 Midnight	8.0	9 p.m.	11.0
1 a.m.	5.0	11 p.m.	8.0	8 p.m.	12.0
18/4 12 Midnight	5.0	10 p.m.	9.0	7 p.m.	13.0
11 p.m.	6.0	9 p.m.	9.0	6 p.m.	14.0
10 p.m.	5.0	8 p.m.	10.0	5 p.m.	15.0
9 p.m.	5.0	7 p.m.	13.0	4 p.m.	16.0
8 p.m.	5.0	6 p.m.	15.0	3 p.m.	16.0
7 p.m.	5.0	5 p.m.	15.0	2 p.m.	15.0
6 p.m.	6.0	4 p.m.	15.0	1 p.m.	16.0
5 p.m.	6.0	3 p.m.	14.0	12 Midday	15.0
4 p.m.	7.0	2 p.m.	15.0	11 a.m.	14.0
3 p.m.	9.0	1 p.m.	15.0	10 a.m.	13.0
2 p.m.	10.0	12 Midday	13.0	9 a.m.	12.0
1 p.m.	9.0	11 a.m.	10.0	8 a.m.	10.0
12 Midday	8.0	10 a.m.	10.0	7 a.m.	10.0
11 a.m.	8.0	9 a.m.	8.0	6 a.m.	10.0
10 a.m.	7.0	8 a.m.	6.0	5 a.m.	9.0
9 a.m.	6.0	7 a.m.	6.0	4 a.m.	9.0
8 a.m.	6.0	6 a.m.	7.0		

7.8 Further reading

Archer M. S. 2004. Weather station ambient temperature corrections. In Forensic Entomology XXII, International Congress of Entomology, 15–21 August 2004, Brisbane, Australia.

Briere J.-F., Pracros P., Le Roux A.-Y. and Pierre J.-S. 1999. A novel rate model of temperature-dependent development for arthropods. *Environmental Entomology* **28**(1): 22–29.

Dillon L. and Anderson G. S. 1996. Forensic entomology: the use of insects in death investigations to determine elapsed time since death in interior and northern British Columbia Regions. Technical report TR-03-96. Canadian Police Research Centre: Ottawa, Ontario.

Grassberger M. and Reiter C. 2001. Effect of temperature on *Lucilia sericata* (Diptera: Calliphoridae) development with special reference to the isomegalen- and isomorphen diagram. *Forensic Science International* **120**(12): 32–36.

Grassberger M. and Reiter C. 2002. Effect of temperature on development of the forensically important holarctic blowfly *Protophormia terraenovae* (Robineau-Desvoidy) (Diptera: Calliphoridae). *Forensic Science International* **128**: 177–182.

Hwang C. and Turner B. D. 2005. Spatial and temporal variability of necrophagous Diptera from urban and rural areas. *Medical and Veterinary Entomology* **19**: 379–391.

Nishida K. 1984. Experimental studies on the estimation of post-mortem intervals by means of fly larvae infesting human cadavers. *Japanese Journal of Forensic Medicine* **38**(1): 24–41 [English abstract available].

Wilson L.T. and Barnett W. W. 1983. Degree–days: an aid in crop and pest management. *California Agriculture* **January–February**: 4–7.

8

Ecology of forensically important flies

Flies that are attracted to a corpse are influenced by their environment. This information is of vital importance when interpreting a crime scene and estimating the length of time that a body has been dead. Female insects choose to lay eggs in places on a body that provide sufficient food for the new generation, along with protection, moisture and a consistent microclimate for larval development. As it decomposes, the human corpse is large enough to support colonies of a number of different species of fly. Work by Archer and Elgar (2003) showed that after being exposed outdoors for 24 hours, the preferred colonization sites on a carcass changed from the orifices to skin folds. Such locations included between the legs or under the ear pinnae. Dear (1978) summarized the attractants which led successions of flies to oviposit on decomposing bodies (Table 8.1). Archer and Elgar also found that, over time, larvae of both the Calliphoridae and the Sarcophagidae altered their distribution on the body, and concluded that migration to more favoured sites was in response to food depletion.

8.1 Ecological features of bluebottles (Calliphoridae)

Blowflies are amongst the first to colonize a body. In the UK, possibly the most common species initially present on the body are *Calliphora vicina* and *Calliphora vomitoria*. One of the first clues to their colonization is the presence of eggs on the corpse. Campan *et al.* (1994) used a lexical software package to analyse verbal descriptions of the sexual behaviour of *Calliphora vomitoria* in order for mating to be successful and oviposition to be possible. Their work revealed the need for correct orientation of the body of the fly and presentation of its wings.

Mating behaviour most frequently leads to flies laying eggs on a body during daylight, although some researchers have also found that oviposition in some species can take place at night. Wyss *et al.* (2003b) recorded flies laying eggs up until 10 p.m., unless it was raining. Greenberg (1990) also recorded egg laying at night, although in his experiments the bodies were kept at ground level. Singh and Bharti (2001) investigated night time oviposition and used bait placed 1.85 metres (6 feet) above the ground, on top of a pole, to counter the argument that the flies had crawled to the copse to lay eggs during the dark. They used suitably sticky tape

Table 8.1 Succession of Diptera on carrion in the open in the UK

Wave of colonization	Families and species	Trigger for colonization	Comments
1	*Calliphora* sp. *Lucilia* sp.	Blood, faeces, urine, initial death odour	*Lucilia* spp. are thought to attract *Calliphora subalpina*, *Calliphora lowei* and *Calliphora uralensis* *Musca* sp. *and Muscina*, if attracted, do not oviposit
2	*Lucilia* sp. *Calliphora* sp. Muscidae Fanniidae Piophilidae Sphaeroceridae	Body gases; the corpse becomes bloated; active decay starts	*Cynomya* may dominate in the north, but *Lucilia* spp. and *Calliphora* spp. remain the most common in other areas Spharoceridaea may include *Graphomya* sp. and *Hydrotea* sp. *Piophila latipes* and *Piophila varipes* are the most common piophilids Anthomyids may be feeding on orifice exudates
3	*Fannia* sp., Stratomyidae, Piophilidae and *Neoleria* sp. (Heleomyzidae) present as larvae or ovipositing adults	Rancid fats are present	*Calliphora* and *Lucilia* now at the post-feeding stage and leave the body Sphaeroceridae and Platysomatidae will also be present *Piophila casei* tends to be a later piophilid colonizer
4	*Piophila casei*, Sepsidae, Drosophilidae, Sphaeroceridae, Ephydridae and Milichiidae	Butyric fermentation	Cheesy smell prevails and in large bodies, such as humans, *Eristalis*, a syrphid, may be present feeding off the decomposition fluid
5	Coleoptera, Sphaeroceridae, and Phoridae	Ammoniacal fermentation stage	Coleoptera dominate the succession assemblage from this stage. *Hydrotaea* (*Ophyra*) sp. may be attracted to the body. Phorids will invade drying tissue

After Dear (1978), with permission from the Amateur Entomologists' Society.

wound round the pole, to prevent flies crawling rather than flying to the bait. At temperatures of 16–27°C, very low light levels (0.6–0.7 lux) and relative humidities (RH) of 75–85 %, the calliphorids *Calliphora vicina*, *Chrysomya megacephala* and *Chrysoma rufifacies* laid eggs at night, in experiments run in both March and September. These results supported the observations of oviposition at night by *Calliphora vicina*, *Phormia regina* and *Lucilia (Phaenicia) sericata*, at locations lit by street lights, which had been previously published by Greenberg. The author commented that the numbers of eggs laid at night are few. However, other authors do not record the same successful oviposition at night (Greenberg, personal communications; recorded in Greenberg and Kunich, 2002). When investigating a death, therefore, the possibility of oviposition on a body during darkness should not be excluded.

Similarly, weather conditions should also be taken into consideration. Greenberg (1990) stated that calliphorids do not fly in the rain. Digby (1958) examined the effects of wind speed on the ability of *Calliphora vicina* (then known as *Calliphora erythrocephala*) to fly. He recorded an optimal wind speed for flight of 0.7 metres/second and suggested that speeds greater than this inhibited the ability of *Calliphora vicina* to fly. Temperatures above 30°C and below 12°C are also known to inhibit blowfly activities. This fact should be taken into consideration when interpreting the conditions at a crime scene, as the blowflies may not have been in a position to fly to find a body where the temperatures have not been favourable.

Daily activity patterns in blowflies are influenced by seasonality and change with geographic location. Hedström and Nuorteva (1971) considered most calliphorids to have a maximum daily flight activity after midday, or to show bimodal flight activity, although in the sub-arctic regions of Finland peak flight activity was around noon. In central Asia *Calliphora vicina* shows a single (unimodal) peak of daily activity in the cooler months, when the flies are most active around midday. In contrast, in the warmer months in central Asia, this species is active at two periods in the day, with least activity during the hottest part of the day (Erzinçlioğlu, 1996). Changes in peaks of seasonal activity will also be related to seasonal peaks in fly populations. Johnson and Esser (2000) suggest that in the tropics, blowfly population peaks are synchronized with the early and late stages of the rainy season, when relative humidities and temperatures are high but rainfall is not at its maximum. Thus, the influence of the seasons may affect the interpretation of when the eggs were laid on a corpse.

In general, blowflies overwinter in soil as third instar larvae. *Lucilia* spp. have a maternally-induced diapause in the third instar; this differs in *Calliphora vicina*, in which populations only undergo diapause if they are northern, since the weather in the north can drop below freezing. Both *Calliphora vicina* and *Calliphora vomitoria* tolerate a certain degree of supercooling. *Calliphora vicina* has a lower freezing threshold; the egg stage is particularly cold-resistant (Block *et al.*, 1990). Survival strategies of particular species and the location of the corpse should therefore be considered when interpreting the post mortem interval from species which are present on a body early in the year.

The presence of calliphorid flies and their egg-laying activities in the UK in winter has been the topic of a study undertaken in London, during mid-December

(Brandt, 2004). Egg masses were found on pig carcasses in both indoor and outdoor locations in central London, at ambient indoor temperatures of 10–16°C and an ambient outdoor temperature range of –1°C to 14°C. The eggs on the indoor pig hatched into larvae, whilst those on the outdoor pig remained in the egg stage. A similar response to temperature was observed in December in Lincoln, where at 9°C ambient temperature, pig skin exposed on a balcony, as part of a student final year project on finger prints, was used by a visiting fly as an egg-laying site. The eggs were removed and placed in a controlled environment cabinet to hatch and were subsequently found to be *Calliphora vomitoria.*

Climatic change may be playing a role in altering the distribution of Calliphoridae and be responsible for changes in the range of some species. This is particularly true of *Phormia regina,* which is a common species at crime scenes in the USA. This species is attracted to human faeces and animal dung (Coffey, 1966). Byrd and Allen (2001) showed that it is the dominant species in forensic contexts in the summer months in northern USA, whilst being the dominant species in the winter months (October–March) in southern USA. Its activity is inhibited by temperatures as low as 12.5°C, according to research undertaken by Haskell (cited in Byrd and Allen, 2001).

Huijbregts (2004) confirmed the presence of *Phormia regina* in The Netherlands, Fennoscandia and the UK, noting that in the twentieth century it has only been recorded on four occasions. Indeed, Erzinçlioğlu (1996) considered *Phormia regina* to be introduced into the UK from the USA. However, in 2001 Huijbregts (2004) recorded *Phormia regina* in The Netherlands on four separate occasions in the same year. He suggests that this species is extending its range. Catts and Haskell (1990) comment that this fly prefers shade rather than brightly lit, open habitats. This preference for shade may account for the prevalence of records of *Phormia regina* in urban environments in The Netherlands, where shade is readily available in and around buildings.

In laboratory experiments, Byrd and Allen showed that development times for *Phormia regina* were increased under a cyclical temperature regime, compared to development times at a constant temperature. However, using constant temperatures they were able to replicate the results for time to peak egg hatching recorded by Kamal (1958). Peak emergence was found at 19 hours at 25°C and at 15.5 hours at 30°C (Byrd and Allen, 2001). Variation in the length of the life cycle occurred at higher temperatures (35 − 45°C), when adults failed to emerge. They also found a variation in the length of the life cycle when cultures were kept at a constant temperature of 40°C or at 10°C (this observation is interesting as, in *Phormia regina*, activity is thought to be inhibited when monthly temperatures have an average of below 10°C; Deonier, 1942). Byrd and Allen felt that developmental times at experimental temperatures of 25°C and below were in accord with the rest of the scientific literature for this species, but they urged caution in using published data to determine the 'experimental value for accumulated degree hours or days' where environmental temperatures at the crime scene were above 25°C.

A second species which is beginning to be more frequent in Europe, as indicated by Dutch crime scene records, is the holarctic blowfly, *Protophormia terraenovae*. Whilst this species is not often recorded from rural environments, it has become more common in large cities along The Netherlands coast. In Britain it is recorded from the Pennines, and Erzinçlioğlu (2000) states that it favours upland and northern areas. Nuorteva (1963) suggests that this species does not compete well with other species and so only flourishes where competition is low, e.g. as an Arctic species. He also suggests that it is usually the first blowfly to emerge, and if a larva of this species is found on a body, then it indicates that the death occurred in the spring.

Changes in distribution are also occurring in *Chrysomya* spp. that are considered to be tropical and sub-tropical. They are found in Africa, Asia and southern Europe and were introduced onto the American continent. Competition with *Chrysomya albiceps* is now thought to be causing a reduction in the prevalence of the secondary screwworm fly, *Cochliomyia macellaria*, a species which is native to the American continent (Faria Del Bianco *et al.*, 1999). Second and third instar larvae of *Chrysomya albiceps* are predators on other larvae when the occasion presents itself (this is termed being a *facultative* predator), whilst first instar *Chrysomya albiceps* larvae feed on tissue fluid or decomposition liquor. This species cannot complete its development below 15°C (unless it is present in sufficient numbers that it can form a maggot mass, which in itself raises the local temperature of the microclimate for the larvae). The time between oviposition and the adult stage is recorded as 19.2 ± 0.92 days at 20°C and 8.3 days ± 0.5 at 35°C (Grassberger *et al.*, 2003).

In Afro-tropical regions, oriental regions from India to China, central South America and southern Europe, *Chrysomya albiceps* is commonly the initial colonizer of a corpse (Baumgartner and Greenberg, 1984; Hall and Smith, 1993). It is also recorded as one of the two most encountered species in forensic cases in South Africa (Mostovski and Mansell, 2004), where it is recognized as a spring and summer species. Smith (1986) considered that this species fulfils the initial colonizing role undertaken by *Calliphora* and *Lucilia* in the temperate zone. However, *Chrysomya* spp. have also been recorded from northern France (Erzinçlioğlu, 2000). Its range seems to be moving northwards and in 2002, *Chrysomya albiceps* was recorded from Austria in central Europe (Grassberger *et al.*, 2003). The level of interaction of this species with other species, rather than the influence of higher temperatures alone, is considered by Grassberger *et al.* to be the important factor in determining its change in distribution.

A species recorded in Britain as colonizing a body in the second wave of insects is *Cynomya mortuorum* (L.), a blue-green blowfly (Plate 2.17). Under Finnish field conditions, its life cycle from egg to adult takes on average 26.2 days at an average temperature of 15°C (Nuorteva, 1977). Staerkeby (2001) recovered *Cynomya mortuorum* from a suicide victim in south-eastern Norway and concluded that its egg-laying seasonality in Norway was similar to that in Finland. In the UK this species is more common in northern England and Scotland than in the south of

England, where it is considered to be scarce (Colyer and Hammond, 1951; Smith, 1986), so the species may be considered to be present in small numbers on a corpse and also to be a geographic indicator for the north of England and Scotland.

8.2 Greenbottles – *Lucilia* spp.

In Britain there are seven species of greenbottle. These are *Lucilia (Phaenicia) sericata*, *Lucilia caesar*, *Lucilia illustris* Meigen, *Lucilia richardsi*, *Lucilia silvarum* Meigen, *Lucilia ampullacea* Villeneuve and *Lucilia bufonivora* Moniez. Smith (1986) records only four of them, *Lucilia ampullacea*, *Lucilia illustris*, *Lucilia caesar* and *Lucilia sericata*, from dead bodies in the UK. They vary in their choice of habitat. *Lucilia* species are amongst the first wave of blowflies to colonize a body. *Lucilia sericata* is, however, reported in experimental trials by Fisher *et al.* (1998) on the basis of poor responses to fresh compared to aged liver, to be a species which does not act exclusively as a pioneer species on a fresh corpse. This may be because this species, when ovipositing, has a preference for corpse surface temperatures above 30°C (Cragg, 1956).

In southern Britain *Lucilia sericata* has been estimated to produce three or four generations a year (Wall *et al.*, 1993). *Lucilia sericata*, which is very infrequently found indoors, or in woodlands and hedgerows (Smith and Wall, 1997), is thought, in Britain, to be a species indicative of bright sunlight (Colyer and Hammond, 1951). In the USA, this preference for sunlight enables *Lucilia* (syn. *Phaenicia*) *sericata* to colonize bodies during the hottest part of the summer, although *Lucilia illustris* is a fly recognized in the USA as associated with those bodies found in open, brightly lit habitats (Catts and Haskell, 1990). In the UK, *Lucilia illustris* is considered to be an open woodland and meadow species and *Lucilia caesar* to be a species associated with forests and tolerant of shade. The latter species occurs further to the north of Britain than *Lucilia illustris* (Smith, 1986).

8.2.1 Ecological indicators

The Finnish community of corpse-inhabiting species in open habitats is bigger, and five species of *Lucilia* are recorded. The species are *Lucilia sericata*, *Lucilia caesar*, *Lucilia illustris*, *Lucilia silvarum* and *Lucilia richardsi*. Of these five, Prinkkilä and Hanski (1995) report *Lucilia illustris* to be the most abundant, followed by *Lucilia silvarum*. Hanski (1987) noted that *Lucilia illustris* was the earlier of the two species to emerge in early summer. Prinkkilä and Hanski considered that *Lucilia caesar*, *Lucilia sericata* and *Lucilia richardsi* are rare, citing their larval competitive abilities as the reason. This poor competition results in varying levels of fly densities in the field.

In general, field populations of greenbottles appear to be small. Gilmour *et al.* (1946) suggest that *Lucilia cuprina* Weidmann, a species found on carrion in

Australia and as an agent of myiasis, has a population density of 0.17–14 flies per hectare. In 1960, MacLeod and Donnelly estimated a northern British population of *Lucilia sericata* to be 2.5 flies per hectare, which was considerably smaller. However, Smith and Wall (1998) showed that the density of *Lucilia sericata* varied within the season, rising from 0 per hectare in England in early June of 1994, when the study took place, to 6.3 per hectare in the middle of August. Clearly, populations will differ with season and geographic location, which may account for their absence from a corpse on occasions when they might be expected.

The distance over which greenbottles disperse also varies with species. In experimental trials in Australia, *Lucilia cuprina* has been recorded 0.7–3.5 kilometres from the point of release (Gilmour *et al.*, 1946), whilst work on the dispersal of individual *Lucilia sericata* suggests that it might be expected to disperse a much smaller average distance of up to 800 metres per lifetime (Smith and Wall, 1998). However the fecundity of the flies and their physiological state can also influence their dispersal. Smith and Wall point out that movement in response to carrion odour is a trigger in both gravid females and those females which lack a supply of protein in their diet, whereas those females which have just laid eggs, or who are newly merged and are undergoing egg development (*vitellogenesis*), show less tendency to respond to the odours emitted from a corpse. The researchers suggest that availability of both sugars and protein sources may account for the differences in migration rates, and therefore hypothesize that a population of *Lucilia sericata* could spread at a rate of 31–42 kilometres per year.

Once the eggs were laid and the larvae emerged, Rankin and Bates (2004) demonstrated that, in cohorts of *Lucilia sericata* above a population size of 400 larvae, heat production was dependent on which instar was present, and not on the size of maggot mass or cohort. They concluded that at ambient temperatures above 15°C, the duration of the life cycle (and hence post mortem interval) could be based on the ambient temperature of the environment for this species. They also showed that at temperatures of 22–35°C larval growth rates in the wild did not significantly differ from those in laboratory experiments. This point is important when attempting to estimate time since death, on the basis of larval development in controlled environments for this species.

Family Sarcophagidae

This family have been found, both in Britain and elsewhere, to be present on the body after a few days of decomposition. One of the most significant things about this family is that the flies are *viviparous*, i.e. they generally lay larvae onto the corpse and not eggs (although egg-laying in Sarcophagidae has been noted under laboratory conditions).

Sarcophagidae are considered to be unimpeded by rain and to fly despite the weather (Erzinçlioğlu, 2000). As a result, flesh flies may be the initial colonizers of the body if there is a long period of rainy weather. Smith (1986) also stated that flesh flies prefer sunlight rather than shaded conditions. The research undertaken by Kamal (1958) provides an indication of the length of the life cycle stages for the American sarcophagid species *Sarcophaga cooleyi* Parker, *Sarcophaga shermani* Parker and *Sarcophaga bullata* Parker. However, one of the most common species of Sarcophagidae recovered as larvae from indoor crime scenes in the USA is *Sarcophaga haemorrhoidalis* (Fallen) (Byrd and Butler, 1998).

Family Phoridae

This family of flies are particularly associated with the final stages of decomposition, where Diptera and Coleoptera are both found on the remains of a corpse, located on the surface of the ground and in advanced decay. However, one species is particularly important if buried bodies are being investigated. This is *Conicera tibialis*, the coffin fly. It is specialized at feeding on this food source and the adult females appear to be able to locate a body using odours emitted from the ground. There is evidence that this species can complete several generations below ground without emerging from the soil, since puparia of *Conicera* sp. have been recorded from exhumed human corpses (Colyer and Hammond, 1951).

Examples of other phorid flies that have been found in forensic cases are *Megaselia rufipes* Meigen, *Megaselia scalaris* Loew (a parasitic fly) and *Triphleba hyalinata* Meigen (a cave dweller). Of these, *Megaselia rufipes* is the most commonly encountered species of phorid on corpses left on the soil surface or in shallow graves in Britain (Disney and Manlove, 2005).

Disney (2005), working on *Megaselia giraudii* (Egger) and *Megaselia rufipes*, has investigated the life cycles of these scuttle flies to enable them to be used for post mortem interval determination. He found a large degree of variation in larval growth from the same batch of eggs and noted that successive waves of larvae migrated from their feeding site.

8.3 Ecological associations with living organisms

Associations of insects with living species, where the living body becomes the food source for fly species, is termed myiasis. The incidence of myiasis is known worldwide. One of the most well researched agents is *Lucilia sericata*, a source of myiasis in sheep in Australia and in the UK. In Africa and oriental environments, *Chrysoma megacephala* and *Cochliomyia macellaria* are notable species colonizing living flesh.

In the UK, the most common form of human myiasis is enteric myiasis. Smith (1986) indicates that the larvae of the following genera have been identified as the cause of this disease: *Eristalis*, *Piophila*, *Drosophila*, *Calliphora* or *Musca*. Where urinogenital myiasis has been diagnosed, the species *Fannia canicularis* (the lesser house fly), *Musca domestica* (the house fly) and *Sylvicola fenestralis* (Scopoli) have been the culprits.

Colonization of flesh by *Lucilia caesar* and *Protophormia terraenovae* is commonly related to latitude. In the case of both of these species, their status as agents of myiasis becomes increasingly more common the further north one goes (Stevens, 2003). In countries such as Finland, *Protophormia terraenovae* is a predominant cause of sheep strike.

8.3.1 Piophilidae

Mullen and Durden (2002) point out that *Piophila casei*, the cheese skipper, will colonize a dead body where the bigger calliphorid and sarcophagid species are prevented from reaching the body, However this species has also been found to colonize living bodies too. Because these insects are found inhabiting both cured ham and cheese they have, in the past, been found to have been consumed accidentally. James (1947) reports the existence of frequent cases of human intestinal myiasis as a result of such consumption of *Piophila casei*. These days this is less likely to happen because hygiene laws and packaging techniques mean that the consumer predominantly eats cook–chill or ready meals.

There is evidence that dogs can consume the larvae of *Piophila casei* and that the larvae do not die as they pass through the canine intestinal tract (James, 1947), although this conclusion was not substantiated by an experiment described in volume 12(3) of the *Bulletin of Entomological Research*, undertaken by Patton (1922), who fed a dog eggs and larvae of myiasis-causing Diptera when working in India. In this instance, both life stages appear to have been digested. In contrast, Austen (1910; cited in James, 1947) recorded *Piophila casei* as causing nasal myiasis in a patient who had extreme nasal discharge and complained of pain over a number of weeks. Larvae expelled from the nose were reared to the adult stage and *Piophila casei* was confirmed as the cause.

8.3.2 Sepsidae

Sepsid flies are common on bodies at a stage of advanced putrefaction. These are a family of flies which are attracted to decaying material and so justify their common name of scavenger flies. The larvae are found in decaying organic matter, particularly when it is liquefying (Oldroyd, 1964), including the decay stages of decomposition of dead bodies. The adult flies have a characteristic wing-waving movement and, according to Oldroyd (1964), will run up and down on any available surface or vegetation.

Smith (1986) places them between the caseic decomposition stage and before the ammoniacal fermentation of advanced decay, in what Mégnin, updated by Smith (1975), called the fourth wave of insect succession. Smith associates them with

Piophilidae and Drosophilidae and identified seven species specifically associated with carrion or from human faeces. Because this family is associated with dung, its presence on a corpse requires careful consideration. Does it indicate faecal contamination of the body in the process of death, or does their presence reflect environmental conditions which are unrelated to the presence of the body? This was a point made quite strongly by Smith (1986).

8.3.3 Trichoceridae

The winter gnats have been recognized as a family within which species could be used to determine post mortem interval in humans (Broadhead, 1980; Erzinçlioğlu, 1980, 2000), although they have not yet been recorded from many human corpses. In fact, Trichoceridae are usually found in rotting vegetation, and have been found in such places as rotting potato clamps (Keilin and Tate, 1940). In volume 12 of the *British Checklists of Diptera*, nine species of *Trichocera* are listed, with two more known but not yet identified. Particular species of winter gnat may potentially be of value on a local basis. For example, in the north of England Erzinçlioğlu recovered the species *Trichocera annulata* Meigen from an ox heart, in an experiment on colonization of flesh which he ran in his garden between 8 December 1979 and 1 March 1980.

Although present throughout the year, the winter gnats are most obvious in winter and can be seen flying even when snow is on the ground, or in late afternoon in winter (Colyer and Hammond, 1951). However, they are more apparent and more abundant in spring and autumn than in summer. Winter gnats can be seen in swarms, when males are moving together as a group, in what has been termed 'a dance'. Edwards (1928; recorded by Keilin and Tate, 1940) suggests that females remain in sheltered roosting sites, and that any that do enter the swarm of males will immediately undergo copulation.

In *Maggots, Murder and Men* (2000), Erzinçlioğlu describes a case where *Trichocera* larvae were found on the corpse of a 9 year-old girl at a similar time of year to that of his experiment. The girl, Zoe Evans, was subsequently found to have been murdered by her stepfather, Mike Evans, a private in the armed forces. The murder had taken place in January and 6 weeks later, in February, the decomposing body of the girl was found wedged head-first into a badger's sett. The girl had died by suffocation, from a combination of having had her shirt stuffed into her mouth and inhaling blood. The blood had come from her receiving a violent blow, sufficient to give her a broken nose. The police confirmed their suspicions as to the murderer when they found some of Mike Evan's clothes and underwear that were covered in blood. The only significant evidence of an entomological nature found on Zoe Evan's body were larvae of Trichoceridae, which confirmed that the murder had taken place in winter, as everyone knew. In this instance, the murder provided more information about the insects than did the insects about the murder, and confirmed that these are a family which could be used to determine time since death in the winter!

8.4 Further reading

Bonduriansky R. and Brookes R. J. 1999. Reproductive allocation and reproductive ecology of seven species of Diptera. *Ecological Entomology* **24**: 389–395.

Disney R. H. L. 1973. Diptera and Lepidoptera reared from dead shrews in Yorkshire. *The Naturalist* **927**: 136.

Disney R. H. L. 1981. A survey of the scuttle flies (Diptera: Phoridae) of upland habitats in northern England. *The Naturalist* **106**: 53–66.

Erzinçlioğlu Y. Z. 1980. On the role of *Trichocera* larvae (Diptera: Trichoceridae) in the decomposition of carrion in winter. *The Naturalist* **105**: 133–134.

Erzinçlioğlu Y. Z. 2000. *Maggots, Murder and Men.* Harley Books: Colchester.

Faria Del Bianco L. and Godoy W. A. C. 2001. Prey choice by facultative predator larvae of *Chrysomya albiceps* (Diptera: Calliphoridae). *Memoirs of the Institute Oswaldo Cruz* **96**(6): 875–878.

Gilbert M. G. and Bass W. M. 1967. Seasonal dating of burials from the presence of fly pupae. *American Entomologist* **32**: 534–535.

Hayes E. J., Wall R. and Smith K. E. 1999. Mortality rate, reproductive output and trap response in populations of the blowfly *Lucilia sericata*. *Ecological Entomology* **24**: 300–307.

Pont A. C. and Meier R. 2002. The Sepsidae (Diptera) of Europe. *Fauna Entomologica Scandinavica* **37.**

Tabor K. L., Fell R. D., Brewster C. C., Pelzer K. and Behonick G. S. 2005. Effects of ante mortem ingestion of ethanol on insect successional patterns and development of *Phormia regina* (Diptera: Calliphoridae). *Journal of Medical Entomology* **2**(3): 481–489.

Vinogradova E. B. 1984. The blowfly, *Calliphora vicina*, as a model object for ecological and physiological studies. *Proceedings of the Zoological Institute, Russian Academy of Sciences* **118**: 272. Nauka: Leningrad.

9
Ecology of selected forensically important beetles

Ecology is the study of the relationship between organisms, in this case beetles, and their environment. For forensically important species this environment is the crime scene, which may include a corpse. Over time the nature of the corpse changes as it decomposes. The insects attracted to a corpse not only utilize it as a food source and habitat but also change its attractiveness to particular species. As a result, the dominance of populations of particular species alters and a succession of insects is recognized. This sequence is termed *insect succession*. It is this sequence of change of insect species that can be used by forensic entomologists to estimate how long the person has probably been dead.

Payne (1965) was amongst the first scientists to design an experiment to relate the stages of body decomposition to insect succession. He also characterized the feeding style of the species and separated those which actively fed on the corpse from those who were 'just passing' and those who were predators on the original specimens.

9.1 Categories of feeding relationship on a corpse

Four main categories of feeding relationship have been described (Campobasso *et al.*, 2001). These are:

- *Necrophages*, which feed only on the decomposing tissue of the body or body parts, e.g. Nitidulidae and Dermestidae.

- *Predators* (and *parasites*) of the necrophages, e.g. Staphylinidae.

- *Omnivores*, which consume both the live specimens inhabiting the corpse and the dead flesh, e.g. ants (Formicidae).

- *Opportunist species* (*adventitious*), which arrive because the corpse is a part of their local environment, e.g. mites (Acari), butterflies (Lepidoptera) and on occasion spiders.

Rodriguez and Bass (1983) showed that information about succession in relation to decomposition could be used to determine the post mortem interval of human corpses. This method of applying succession to determining the post mortem interval is based upon knowledge of the local fauna. It may also require experimentation to confirm the sequence of colonization in a particular location (Table 9.1). For example, if the species present included X, Y and Z and these particular species had been shown to be present in the locality 14–16 weeks after invading a fresh corpse, then time since death of that person would be estimated as 14–16 weeks. Such an assemblage of insects would define the 'probable' time since death and would be a guide to post mortem interval. In situations where the body is badly decomposed and the forensic pathologist cannot provide an estimation of time of death, using information from the succession of insects may be the best estimate that is available, despite the large margin of error which is ascribed when interpreting the data.

Schoenly and Reid (1987) questioned the view that insect succession could reliably distinguish a time period for post mortem interval. Based upon 11 studies of insect succession, they found that the timing of the succession varied and proposed insect succession on the corpse to be a continuum. However, clear relationships between decomposition states, habitat conditions and the presence and sequence of beetle families have been demonstrated. For example, Oliva (2001) showed that in Argentina the nitidulid beetle *Carpophilus hemipterus* (Linnaeus) was found in the later stages of decomposition, in association with the Piophilidae and often also with *Necrobia rufipes*, a clerid beetle. Oliva also linked silphids of the genus *Hyponecrodes*, such as *Hyponecrodes erythrura* Blanchard, with corpses recovered from rural outdoor environments. However, caution has to be used in interpreting scientific data from a crime scene in one location, or country, to another. Ideally, data about insect succession on carrion for the particular region where the death took place should be used.

Table 9.1 Generalized coleopteran succession on carrion in relation to stage of decomposition. Only species from those families which are found breeding upon the corpse can truly be considered as forensic indicators

	Fresh	Bloat	Active decay	Advanced decay	Skeletal
Staphylinidae	*	*	*	*	*
Histeridae			*	*	*
Cleridae		*	*	*	
Scarabaeidae		*	*		
Carabidae			*		
Silphidae		*	*	*	
Dermestidae			*	*	*
Nitidulidae				*	*
Trogidae				*	*
Geotrupidae			*		
Tenebrionidae					*

In North American studies, the first families of beetles recorded on the body are the carrion beetles (Silphidae), the rove beetles (Staphylinidae) and the clown beetles (Histeridae) (Anderson and VanLaerhoven, 1996; VanLaerhoven and Anderson, 1996). Amongst the later colonizers are the dermestids. Mégnin (1887) observed that hide beetles (Dermestidae) were attracted to a body between the third and the sixth month after death, when it was in advanced decay. At this stage, body fat has decomposed and butyric acid becomes a predominant component of the odour plume arising from the body.

The speed of decomposition is influenced by environmental conditions, so dermestids may also be found sooner. In Canada, VanLaerhoven and Anderson (1996, 1999) recorded them some 21 days after death, when the body was in early advanced decay. Oliva (2001), working in Argentina, also found early colonization of bodies by dermestids, 10–30 days after death. Seasonality, temperature of the environment, humidity, level and duration of rainfall and insect abundance in the locality are all major influences on the rate of decomposition of the body and therefore on the speed of succession of the insects colonizing it.

Work relating to insect succession on rabbit corpses in Egypt revealed rapid decomposition to the dry stage in 4.5 days, at an average dry season temperature of 28°C. This changed for temperatures in the range 13.5–16.6°C, when decomposition to the same point took an average of 51.5 days (Tantawi *et al.*, 1996). In 'autumn', Tantawi recorded a slower rate of decomposition, resulting in a longer period of decay, than in the cooler temperatures of 'winter'. This reduction in the speed of decomposition was thought to be the effect of rainfall delaying larval development. So it is necessary to take weather conditions into account when using succession as a measure of post mortem interval, just as it is when using larval growth rates (accumulated degree hours, ADH).

Insect succession on buried remains is more restricted than on bodies left on the soil surface. The investigation of buried corpses requires a greater investment of resources and time. According to VanLaerhoven and Anderson (1996, 1999), insect succession on buried remains was first studied in Canada in 1995. They considered that before this date no scientifically valid work, simulating disposal of a murder victim, had been undertaken. Their work on insect succession on buried, clothed pigs showed that the species range, measured as colonizers and/or trapped in pitfall traps, was less on previously exposed pigs than on those which had been immediately buried. They noted that buried pigs showed a distinct pattern of succession which contrasted with that from pigs retained on the ground surface. However, they also noted variation, in terms of the species colonizing the body and times of colonization, between the two sites they chose.

This research highlights how important it is to know as much as possible about the species of insects which colonize a body and what influences their development on the corpse. The ecology of selected families and species which colonize the body is presented below.

9.2 Ecology of carrion beetles (Silphidae)

Silphidae, comprising two subfamilies (Lawrence and Newton, 1982), carrion beetles (Silphinae) and burying beetles (Nicrophorinae), frequently colonize a body in the decay stages of decomposition. However, in some countries, species of silphid are notable forensic indicators of fresh and bloated corpses. *Oxyletrum discicollis* Brullé is recognized in this role in south-eastern Brazil and comprised 8 % of the insect assemblage infesting human corpses in urban and rural areas near Cali, Columbia (figures based on 12 deaths; Barreto *et al.*, 2002). They investigated 16 corpses, two women and 14 men, brought to the Cali Institute of Legal Medicine from both urban and rural murders; 75 % of these bodies were infested with adult beetles of this species. In contrast, Wolff *et al.* (2001) found the same silphid species present on pig carcasses later in the decomposition sequence. They recorded it in the active decay stage, at 1450 metres above sea level and a temperature range of 18–24°C.

The particular species of Silphidae present may be influenced by the size of the corpse. Silphinae tend to colonize larger corpses, e.g. Payne (1965) recorded *Necrophila* (= *Silpha*) *americana* Linnaeus at the active decay stage on a large corpse, whereas Nicrophorinae (burying beetles), such as *Nicrophorus humator*, were commonly found on small carcasses of voles and mice. Because much research has been carried out on the Nicrophorinae, they serve, despite their preference for small corpses, to illustrate the ecology of members of the family as a whole.

These beetles exhibit communication between individuals and social behaviour, and many look after their young. Sex attractants (pheromones) have, for example, been postulated for *Nicrophorus vespilloides*. Bartlett (1987) showed that in the laboratory females were significantly more likely to be attracted to containers with males than those containers without, suggesting that an odour might be released by males. Other males were also more likely to be attracted to a location where a male beetle was already present; where the carcass is large, broods can be reared by more than one mating pair. Indeed *Nicrophorus* sp. males may 'call' when they have found carrion. Bartlett recorded signalling ('*sterzeln*'), which was achieved by the tip of the male's abdomen quivering and by it being stroked with the beetle's hind legs. Such communication may lead to several individuals initially locating a carcass. Eventually the corpse usually becomes the habitat for a single pair of beetles, because the individuals fight each other (Pukowski, 1933; in Bartlett, 1987; Wilson *et al.*, 1984). Males and females jointly build a chamber below ground and alter the carcass to support the development of their offspring. Members of the subfamily Nicrophorinae (burying beetles) have been shown to slow down carcass decomposition by producing secretions that are inhibitors of bacterial growth (Hoback *et al.*, 2004). In general, such inhibitors have not been shown for members of the other subfamily (Silphinae).

The location of a suitable carcass triggers female development in some species. Wilson and Knollenberg (in Wilson and Fudge, 1984) discovered that female

Nicrophorus orbicollis Say had underdeveloped ovaries until the beetles encountered an appropriately decomposed carcass. Once this had been located, the females produced eggs over a 48 hour period (Wilson and Fudge, 1984). The numbers of eggs varied with species. *Nicrophorus defodiens* Mannerheim produced an average brood size of 23.9 eggs, whereas *Nicrophorus orbicollis* produced a smaller average number of 14.9 eggs (Wilson and Fudge, 1984).

In some species only the female stays with the larvae until they pupate, whilst in others, such as *Nicrophorus vespilloides*, both males and females may stay (Wilson and Fudge, 1984). The duration of this parenting role is quite short. In their field studies during May to August, Wilson and Fudge showed that the time between adults burying a carcass and the offspring reaching the pre-pupal stage was 10 days. The value of having both parents present is that they can defend the corpse against competition, including preventing flies laying eggs.

There are differences in the distributions of species of Silphidae. *Nicrophorus* species, other than *Nicrophorus vespillo*, are common in forests. Work by Ruzicka (1994) showed that *Nicrophorus vespillo* was more common in fields. Pukowski (1933) considered that the nature of the soil could account for the difference in distribution of some species. Investigations of *Nicrophorus vespilloides* and *Nicrophorus humator* around Frankfurt, Germany, revealed that the former favoured drier soil. Therefore, the presence of a particular species of silphid beetle may be determined by the condition of the carcass, the population of members of that species present in the locality and also the environmental conditions.

9.3 Ecology of skin, hide and larder beetles (Dermestidae)

Several species of dermestid beetles have been shown to colonize a dead body. These include *Dermestes ater* DeGeer, *Dermestes maculatus*, *Dermestes lardarius* and *Dermestes frischii* (Kugelann) (Centeno *et al.*, 2002). *Dermestes maculatus* will be used as an example of the response of dermestids to a corpse, since this species is well researched because of its role as a stored product pest.

Dermestes maculatus growth from egg to adult can take 20–45 days, although the speed of development depends on the temperature of the habitat. The larvae have characteristic hairs on their body segments and are referred to colloquially as 'woolly bears'. These hairs occur in tufts at the end of the body or along the sides of each segment and, according to Hinton (1945), can be moved or vibrated when the larva is being threatened.

The nature of the food is important to the success of dermestid colonization. McManus (1974) considered that the optimum rate of energy consumption for *Dermestes maculatus* was 0.17–0.28 kilocalories per gram per day. Where *Dermestes maculatus* was raised on fish with a high lipid content as a food source, a shorter length of larval stage was recorded (Obsuji, 1975). *Dermestes* spp. have been shown to require dietary sterols, including cholesterol, campesterol or 7-dehydrocholesterol to complete their life cycle (Levinson, 1962). Once the larvae have reached the prepupal condition, they migrate to pupate. This can result in larvae boring into a variety of substances in order to avoid cannibalism as they pupate (Figure 9.1). In addition, Dermestid larvae can delay the time of their

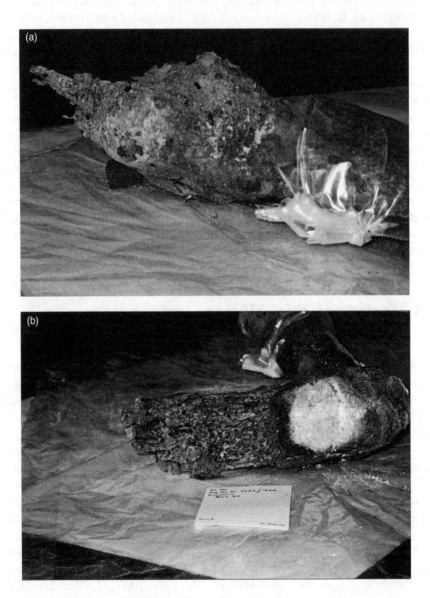

Figure 9.1 Dermestid damage of (a) the leg and (b) the foot of a 'mummy'. Reproduced with permission from Bristol Museums, Galleries and Archives, Bristol City Council

pupation by up to 20 days if there is no suitable place to pupate (Archer and Elgar, 1998).

Adult dermestids show a negative response to light (*negative phototaxis*) and will, when touched, readily 'play dead' (show *thanatosis*). Dermestids will happily exist in darkness as larvae. However, when food is in short supply, the beetles have been known to walk or fly away from the current food source towards a light source. These habits mean that they can be kept in the dark but need a reliable food source and pupation site to successfully complete their life cycles.

On a body on which no live specimens of dermestids remain, their *frass* provides evidence of forensic significance; being an indicator that this species was formerly present. Frass has a characteristic twisted shape and is white in colour (Figure 9.2). It comprises undigested food, which is encased in peritrophic membrane. Where frass alone is present, it may reflect dermestid activity for a period of time between 1 month and 10 years. Indeed, Catts and Haskell (1990) recorded frass originating from dermestids on 10 year-old mummified bodies retained in a house by a solicitous, but criminally culpable, relative.

Ecological conditions appear to determine whether dermestid species will be present. Arnoldos *et al.* (2005) showed that the coleopteran profile in south-eastern Spain varied in both distribution and abundance throughout the year. They recorded few dermestid species in the earliest stages of corpse decomposition in spring and summer. Subsequently, numbers of the dermestid species increased as the remains

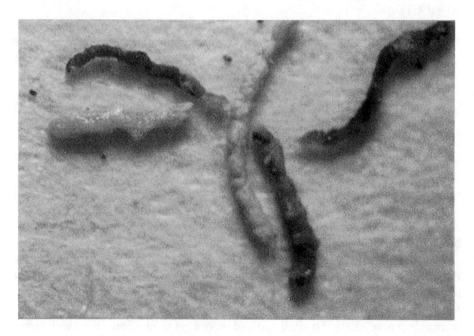

Figure 9.2 Faecal remains (frass) from dermestids (A colour reproduction of this figure can be found in the colour section towards the centre of the book)

began to dry out. Dermestid larvae were characteristic of the dry stage of decay and lots were found in the muscle mass and on bones. In south-eastern Brazil, *Dermestes maculatus* is also recognized as a forensic indicator (Carvalho *et al.*, 2000).

Dermestids appear to tolerate a range of temperatures and relative humidity. Kulshrestha and Satpathy (2001) record dermestids from corpses at an ambient temperature of 16.5°C and 71 % average humidity, but they also noted them on a corpse at an ambient temperature of 20°C and a much reduced average humidity of 46 %. This fits with work by Hinton (1945), who showed that temperatures of 28–30°C resulted in dermestids completing their life cycle in 22 days. At lower temperatures he recorded life cycles of 40–50 days. Raspi and Antonelli (1995) found that the optimum temperature for growth for cultures of dermestids maintained at in constant conditions in a laboratory was 25–30°C, which resulted in an average life cycle duration of 35.1–43.9 days.

Dermestids will compensate behaviourally for high temperatures. During the morning in Nigeria, when carcass internal temperatures were between 24°C and 26°C, *Dermestes maculatus* was seen on its surface (Toye, 1970). Later in the day, when the ambient temperature reached 29–47°C, *Dermestes* moved inside the carcass, where the internal temperature was lower, reaching 29–42°C. The relative humidity within carcasses was found to be 40–70 %. At an experimental temperature of 25 ± 1°C, with two ranges of humidity, one 10–60 % and the other 50–100 %, *Dermestes maculatus* showed a preference for a relative humidity of 50–60 % (Toye, 1970).

As with other carrion-seeking species, dermestids appear to communicate using odour. Conquest (1999) explored the influence of pheromones on the distribution of both male and female dermestids. She rinsed their bodies in the solvent hexane and was able to show attraction of males to both male and female washes. She found that females were attracted to the solution of multiple washings of body odour chemicals in hexane from other females. Males have been shown to secrete a pheromone from a canal beneath the fourth abdominal sternite. Levinson *et al.* (1978, 1981) demonstrated that the most active of these pheromone components were isopropyl Z-9-dodecanoate, isopropyl Z-9-tetradecanoate and isopropyl Z-7-dodecanoate. These chemicals attracted adults and promoted recognition of sexually mature males. Female *Dermestes maculatus* have multiple mates and copulation is achieved more readily with a new mate (Archer and Elgar, 1999). Males show mounting behaviour after copulation and can be found riding on the backs of females, particularly where other males are present.

Jones and Elgar (2004) undertook a laboratory experiment on age-related success in mating in *Dermestes maculatus*. They tested the effect of male age, sperm age and male mating history on female fecundity and their ability to achieve fertilization. They found that where males of intermediate age were used, females were more successfully fertilized and laid more eggs than when mated with either young or older, males. The age of the sperm was not considered an important factor.

The size of a population of *Dermestes maculatus* can also affect the length of the larval period. Both high and low densities increase the length of time of metamorphosis. Rakowski and Cymborowski (1982) suggest that *Dermestes maculatus* produce, and then liberate in faeces, two compounds which influence growth and development. One, produced by the larvae, speeds up growth and encourages aggregation; the other, generated by adults, inhibits larval development. The age of the larvae present on the body should therefore be interpreted on the basis of dermestid population size as well as temperature.

An example of this was provided by Goff (2000), who noted that the last larval skins were shed 51 days after the death of an individual, and commented on the fragility of these freshly-shed dermestid larval cuticles. This fragility, in conjunction with the presence of other species which were also found at similar Hawaiian locations 48–51 days after death, indicated a post mortem interval of greater than this period. However, the freshness of the shed dermestid cuticles on the body, with the absence of larvae, suggested that time since death was not much greater than 51 days. This rapid completion of the larval stage, with the resultant remains of larval cuticle, may have been a response to dermestid population density. Pheromones, where the population of dermestids is large, appear to speed up rates of development, as indicated above. Goff commented that there was forensic significance in the absence of larvae of *Dermestes maculatus* which might be expected to be present in a crime scene in Hawaii.

An association has been found between evidence of the presence of dermestids, with other species, and post mortem interval. For example Arnaldos *et al.* (2005), in their succession studies in south-eastern Spain, recorded Nitidulidae and Dermestidae at the same stage of decomposition, linking their presence on the body. Post mortem interval determination is most accurate when based on evidence of the presence of several species of beetle that are normally found in association, rather than on single species of beetle alone.

9.4 Ecology of clown beetles (Histeridae)

This family is known to be part of the insect assemblage from the bloat stage, through the decay stages and into the dry stage. Histerid larvae and adults feed on the larvae of flies colonizing the body in these decomposition stages. Stevenson and Cocke (2000) explored the life cycle of the histerid beetle *Arcinops pumilo* (Ericson). They suggest that in laboratory-bred cultures, the adult will consume 3–24 muscid eggs per day and that the larva will consume 2–3 eggs per day in order to develop satisfactorily. According to Crowson (1981), at 20–25°C histerid beetles take 31–62 days to pass through their life cycle from egg to adult. The eggs and larvae produced at this temperature tend to be large in size.

Adult histerids have been shown to possess a defensive mechanism and can secrete small drops of a pungent liquid from the ventral surface of their thorax and abdomen. In so doing, they frequently turn over, so their ventral surface is uppermost. Members of this family may also appear, when touched, to be 'dead'. This ability to demonstrate thanatosis is a common defence mechanism found in a number of insect species. This information may assist you in identifying these beetles and not considering them as dead specimens at a crime scene.

Histerids tend to be active during the night and to hide underneath the corpse during daylight. This can account for variations in records of assemblages and the range of species present on the corpse. Equally, the decomposition stage, in which histerid beetles are present on the corpse, can vary from location to location. Korvarik (1995) found that histerid beetles arrived on the body soon after flies had colonized it. This supports the findings of Payne (1965), who recorded them during bloat, which occurred from day 1, in active and advanced decay stages, as well as in the initial dry stages of decomposition, which was recorded from day 5 onwards. Wolf *et al.* (2001), in contrast, recorded the arrival of adult histerids on the corpse 7–12 days after death. They recorded larvae on days 77–118, at the later stages of decomposition.

Richards and Goff (1997), investigating insect succession on pigs placed in woods at different altitudes in Hawaii, recorded *Hister noma* Erichson and *Saprinus lugens* Erichson in their collections. They too stated that histerid beetles invaded a body at the end of the bloat stage. Shubeck (1968) considered that habitat played a large role in determining whether or not histerids were attracted to a corpse. He found that in still air members of this family could perceive odour from sources 1 metre away, but capture–recapture experiments demonstrated little evidence of histerid orientation to bait sources. Little information is available that links the duration of the stages of metamorphosis to histerid colonization of the body at specific temperatures, sufficient to calculate a PMI using members of this family. Determining such information would assist in determining time since death, directly to the presence of histerid species and increase the range of species which can be used for this purpose.

9.5   Ecology of checkered or bone beetles (Cleridae)

Members of the family Cleridae feed on carrion and are often called bone beetles. They have been classified by some workers as members of the Cornetidae rather than the Cleridae, although other researchers have retained the family name Cleridae to include such genus as *Necrobia*. Kulshrestha and Satpathy (2001) comment on the variation in the family names of these beetles. The use of the word 'Cleridae' for the family name has been chosen in this account, as it is a familiar term in forensic entomology.

Cleridae have been found from bloat through to the dry stage of decomposition, although the association with a particular decomposition stage may differ from country to country. For example, in the UK *Necrobia* species can be associated with dry carcasses and bone remains (Cooter, 2006). In India, Kulshrestha and Satpathy (2001) identify Cleridae and Dermestidae as the most common beetles infesting the dry stage of decomposition of human remains. They noted the clerid *Necrobia rufipes* on remains from an environment where the average temperature was 16.5°C and the relative humidity was 71 %, although this species has also been recorded at a higher temperature and a relative humidity of 46 %. This species is called the red-legged bacon beetle, having been a noted stored product pest. It is 4–5 mm long and dark blue in colour. Its legs, and the segments at the base of the antennae, are red.

The biology of *Necrobia rufipes* has been studied experimentally in a light:dark regime of 8:16 hours, at a temperature of 30 ± 0.5°C and a relative humidity of 80 ± 5 %. Bhuiyan and Saifullah (1997) calculated the average number of eggs laid per female to be 89.7 ± 17.8. Around 90 % of these hatched after an average egg stage of 4.1 ± 0.4 days. The length of the larval stage was calculated to be 32.1 ± 5.2 days and that of the pupal stage to be 9.9 ± 1.7 days. The average life span for female *Necrobia rufipes* is 60.6 ± 39.5 days, whilst for males it is shorter (49.4 ± 18.2 days).

Clerids such as *Necrobia rufipes* begin to be attracted at the stage when a body has become saponified and volatile fatty acids and caseic breakdown products are released (Turchetto *et al.*, 2001). Unpublished work by Bovingdon, recorded by Munro (1966), indicated that this species was attracted to copra stored in warehouses, by the stearic and palmitic acids released during fungal growth. It tends to be found on bodies in association with cheese skipper flies (Piophilidae) 3–6 months after an organism dies. Turchetto *et al.* (2001) investigated the association of stage of decomposition with the presence of *Necrobia rufipes*, in the context of a corpse of a young woman in a corn field, who had been strangled. Her badly decomposed body, damaged post mortem by a farm tractor, was found by a hunter in the province of Venice in northern Italy (12 October 1997). *Necrobia rufipes* was found, along with the third larval instar of the piophilid *Steariabia nigriceps* Meigen, a species which is also a member of the British Diptera. Richards and Goff (1997) cite *Necrobia rufipes* as an important forensic indicator species in Hawaii. In Peru, a study of succession of arthropods in relation to body decomposition showed that *Necrobia rufipes* made up 0.45 % of the total insects (4405 specimens) recovered. The study was conducted over 84 days between July and October 2000 (Iannacone, 2003).

In the dry stage, *Necrobia rufipes* and *Dermestes maculatus* can colonize a body at the same time, although their interspecific competition has an effect on population growth of both species (Odeyemi, 1997). At 20°C *Dermestes maculatus* will out-compete *Necrobia rufipes,* whilst at 32°C both species are able to coexist on the same body. This suggests that *Necrobia rufipes* may be at the extreme of its environmental limits and tends to support the previous records of its success earlier in the decomposition sequence.

Clerids may influence interpretation of the cause of death of the corpse. Members of this family, along with silphids and histerids, have been found to cause damage to the cadaver skin and these marks, at first sight, resemble gunshot wounds.

Such holes serve as holes for breeding in, or result from feeding (Benecke, 2004). Therefore, care should be taken in interpreting damage on well decomposed bodies where there is evidence of the presence of members of any of these three families.

9.6   Ecology of rove beetles (Staphylinidae)

Rove beetles arrive on the body in the bloat stage of decomposition, or even sooner. They are predators of fly colonizers feeding on the body and they feed on both the eggs and larvae. Chapman and Sankey (1955) recorded the following species of rove beetle on rabbit carcasses: *Anotylus (= Oxytelus) sculpteratus* Gravenhorst; *Philonthus laminatus* (Creutzer); *Philonthus fuscipennis* (Mannerheim) *Creophilus maxillosus*; *Tachinus rufipes* (Degeer); *Aleochara curtula* (Goeze). These carcasses were placed within 30–40 metres of each other in shrubbery, under a plane tree or in thick meadow grass. Goff and Flynn (1991) recovered specimens of the same genus, *Philonthus* (adult *Philonthus longicornis* Stephens), from samples of sandy soil and leaf litter from beneath where a body had lain at Mokuleia, Oahu, Hawaii.

   The presence of Staphylinidae will vary with the season. In spring, Centeno *et al.* (2002) recorded Staphylinidae on an unsheltered corpse throughout the stages of decomposition. In summer, however, staphylinids were absent from the unsheltered copse and were only recorded in a sheltered corpse during the bloat stage. In contrast, in autumn, on the unsheltered corpse, Staphylinidae were recorded in both the advanced decay and dry stages of decomposition. Their presence cannot be interpreted without considering environmental conditions such as temperature and exposure to sunlight.

9.7   Ecology of dung beetles (Scarabaeidae)

The Scarabaeidae are commonly known as dung beetles. Many of the dung beetle species will inhabit tunnels which they construct beneath a corpse. Two of the most common genera of Scarabaeidae are *Onthophagus* and *Aphodius* (Payne *et al.*, 1968). As with many other beetle species, because they are not immediately obvious on a corpse, their presence can be missed.

   Scarabaeidae, in a study in an urban area conducted in south-eastern Brazil, were the second most frequent colonizer on a pig carcass; the calliphorid *Chrysoma albicepes* was the major colonizer (Carvalho *et al.*, 2000). Three species were considered by Carvalho *et al.* to be important forensic indicators for post mortem determination, because they had been recovered from human cadavers, or from both human cadavers and pig carcasses, in forest environments near Campinas City, Brazil. The species were *Deltochilum brasiliensis* Castelnau, *Eurysternus parallelus* Castelnau, which were found on human cadavers, and *Coprophanaeus* (*Megaphanaeus*) *ensifer* (Germar), which was found with *Canthon* sp. and *Scybalo-canthon* sp. on both pig and human corpses. Despite this association, the presence of suitable food, rather than a particular stage of decomposition, appears to be the deciding factor in whether or not Scarabaeidae are present on a body in any geographic region.

The superfamily Scarabaeoidea includes two other families which are amongst the carrion fauna. These are the Geotrupidae and the Trogidae. Nuorteva collected *Geotrupes stercorosus* Scriba from a partially buried corpse in Finland (Nuorteva, 1977). *Geotrupes* spp. are exclusively dung and carrion feeders. Gill (2005) commented on an association of this family with soil type, noting that Geotrupidae were typically found in areas with sandier soil types. *Geotrupes* spp. presence may therefore be valuable in indicating whether or not a body has been moved, if the soil is not of this type. The second genus of interest to forensic entomologists within the superfamily is the Trogidae, which are carrion feeders at the later stages of decomposition.

9.8 Ecology of trogid beetles (Trogidae)

These insects are found at the dry stage of decomposition. For example, Archer and Elgar (2003) noted that members of several Australian families of beetles, including *Omorgus* sp., a member of the Trogidae, along with *Saprinus* sp. (Histeridae), left fragments of exoskeleton which identified their previous presence. Trogid larvae are easily recognized, if present, as they have a typical 'C' shape. The larvae are reputed to thrive on skin, hair and the remnants of tissue which is dried onto the remaining skeletal bones.

Trogid beetles have been recorded on dry tissue in various seasons. Tabor *et al.* (2004), studying succession on a pig carcass in south-west Virginia, found that Trogidae were spring colonizers of this stage of decomposition. However, in Manitoba, Canada, Gill (2005) recorded *Trox unistratus* Beauvaris throughout the summer, fall and spring periods in her experiments. The family seems to be specific to the later decomposition stage, but not to a particular season.

9.9 Ecology of ground beetles (Carabidae)

Carabids are predators of the insect species which colonize bodies and are most frequently active at night. Larvae of *Nebria*, *Notiophilus*, *Carabus* and *Pterostichus* species are frequently found on the soil surface (Luff, 2006). There are few examples of the contribution of the ground beetles to succession on the corpse and Smith (1986) considered that they were of less importance as predators than other beetle families.

The degree of variation in the assemblage of insects responding to both season and stage of decomposition of the corpse, which has been discussed above, means that only in the context of the local conditions can succession be used to determine the time since death of a corpse at a crime scene. At the crime scene, attention should be given to examining beneath the body for beetles that are nocturnal and hide in soil during the day, as well as to the predatory effects of some families, such as the carabids, which consume eggs or larvae and can cause a gap in the insect profile for a particular stage of succession.

9.10   Review technique: determination of succession and PMI

Investigating the succession of insects on a body requires that you know about the succession of local species and the changes which occur over time, as the nutritional condition of the decaying body changes. From this you can deduce the time period which is most likely to have elapsed between death and discovery of the body.

The practical assignment below is a paper-based exercise, as not everyone has a place where they can carry out practical work on a decomposing body. However, because a choice of meat for human consumption, including whole rabbits, can be fairly readily obtained from a butcher, the practical could be undertaken by using the available information, if you are lucky enough to have a suitable field site and a friendly butcher.

9.10.1   Ecological information about the crime scene

Two volunteers at a field station trapped insects on a daily basis from the carcass of two rabbits, over the first two weeks one July. The bodies of the rabbits had been placed inside a small area of garden shrubbery (Figure 9.3), which was dry and had some leaf litter and sparse vegetation as ground cover.

The species were collected from the carcasses between 8 a.m. and 8.30 a.m. every morning as part of the field station tasks carried out by the volunteers, and are presented as daily catches in Table 9.2.

The samplings of rabbit corpses were repeated for 2 years and no variation in the species present was noted, although the numbers of individuals of the species varied in the second year. In addition to the species listed in the table, specimens of Hymenoptera were also recorded from the body in the second year. On days 12–14 single specimens of the common wasp *Vespula vulgaris* Linnaeus were collected. Five specimens of the braconid wasp *Alysia manducator* Panzer were captured on days 5, 6 and 8. This hymenopteran wasp parasitizes *Lucilia* spp. and *Calliphora* spp. (Smith, 1986).

These data form the background data on insect succession at your site.

Figure 9.3   The shrubbery at the field station where the rabbit carcass was located

Crime scene management information

On 12 July of the year following the second collection, the body of a caucasian male was found under some bushes in a stately home some 5 miles from where the study discussed above was undertaken. The following specimens were collected from the body: *Calliphora vomitoria, Sarcophaga* sp., *Nicrophorus vespillo, Aelochara curtula* and *Hister unicolor.*

- The Officer in Charge is DCS Robert Allen-Hall.
- The Crime Scene number is 01029990705.
- The body was later identified as James Wilson, a local drug user.
- James Wilson was last seen on 3 July.

1. Identify the feeding type of the species listed, to separate those which are forensic indicators from those which are opportunists (adventitionists) or predators. You may have to use references from the Further Reading section to support your investigations. Specify which species are of forensic importance.

2. Draw up a table of 'species versus time' for the presence of the specimens which are of forensic interest.

Table 9.2 Daily field station collections of insect species on a dead rabbit

Order	Species	Day													
		1	2	3	4	5	6	7	8	9	10	11	12	13	14
Diptera	*Calliphora vomitoria*	5	2	5	4	1	1	0	0	0	0	0	0	0	0
	Lucilia sericata	0	0	5	10	4	2	1	5	3	2	1	0	0	0
	Sarcophaga sp.	0	0	3	4	2	1	1	3	0	0	0	0	0	0
	Fannia scalaris	0	0	0	1	1	1	5	6	5	2	1	0	0	0
Coleoptera	*Nicrophorus vespillo*	0	0	0	1	2	2	3	0	0	0	0	0	0	0
	Pterostichus niger	0	0	0	1	0	1	1	0	0	0	0	0	0	0
	Hister unicolor	0	0	0	0	3	4	4	6	1	3	3	0	0	0
	Margarinotus brunneus	0	0	0	0	0	1	0	1	1	0	1	1	2	0
	Saprinus semistriatus	0	0	0	1	6	19	20	20	15	5	20	20	15	12
	Philonthus laminatus	0	0	0	0	0	0	0	0	0	1	1	0	0	0

3. Predict the duration of the most probable post mortem interval in days.

4. Justify your conclusion using the succession table for the location.

9.11 Further reading

Eggert A.-K., Reinking M. and Muller J. K. 1998. Parental care improves offspring survival and growth in burying beetles. *Animal Behaviour* **55**: 97–107.

Forsythe T. G. 1987. *Common Ground Beetles*. Naturalists' Handbooks No. 8. Richmond Publishing: Slough.

Koulianos S. and Schwarz H. H. 2000. Probability of inter- and intraspecific encounters, and duration of parental care in *Nicrophorus investigator* (Coleoptera: Silphidae). *Annals of the Entomological Society of America* **93**(4): 836–840.

Martin L. D. and West D. L. 1995. The recognition and use of dermestid (Insecta: Coleoptera) pupation chambers in paleoecology. *Palaeogeography, Palaeoclimatology, Palaeoecology* **113**(2–4): 303–310.

Moura M. O., Monteiro-Filho de Araújo E. M. and de Carvalho C. J. B. 2005. Heterotrophic succession in carrion arthropod assemblages. *Brazilian Archives of Biology and Technology* **48**(3): 473–482.

Peacock E. 1993. Adults and larvae of hide, larder and carpet beetles and their relatives (Coleoptera: Dermestidae) and of Derodontid beetles (Coleoptera: Derodontidae). *Handbooks for the Identification of British Insects* **5**(3). Royal Entomological Society: London.

Schroeder H., Klotzbach H. and Püschel K. 2003. Insects' colonization of human corpses in warm and cold season. *Legal Medicine* **5**: S372–S374.

Scott M. P. 1998. The ecology and behaviour of burying beetles. *Annual Review of Entomology* **43**: 595–618.

Tullis K. and Goff M. L. 1987. Arthropod succession in exposed carrion in a tropical rainforest on Oahu Island, Hawaii. *Journal of Medical Entomology* **24**(3): 332–339.

Woodruffe G. E. and Coombs C. W. 1979. The development of several species of *Dermestes* (Coleoptera: Dermestidae) on various vegetable foodstuffs. *Journal of Stored Products Research* **15** (3–4): 95–100.

Useful website

A bibliography of papers relating to Dermestidae, including those of forensic interest, is located at: www.dermestidae.com/literat.html

10

The forensic entomologist in court

Forensic scientists of all disciplines, including those studying for forensic science degrees, often develop court room presentation skills through role-play in mock courtroom situations. If you are already employed as a forensic scientist, your employer may train you in this way, as well as by assisting with investigations in the field. You will, over time, gain experience of how others present in court before finally having this responsibility yourself. If you are studying for a degree in forensic science, during your course you will almost certainly be assessed on your written and verbal presentation skills.

In report writing and courtroom skills, good communication is important. This chapter addresses this aspect of study, for those studying for a qualification in forensic science which includes forensic entomology, where you have an opportunity to take part in a mock court. It also provides pointers on writing the report so that you build up your confidence.

The culmination of all of the collecting, identifying and calculating which you have undertaken as a forensic entomologist is the production of a report for either persons or organizations instructing you; in some instances this may be for those seeking redress through the courts. If this report is sufficiently comprehensive and the conclusions clearly stated, you may not need to attend court. Therefore, it pays to learn to communicate the results of an investigation clearly and concisely; to ensure that all of the casework notes are well written and available if requested by the court; accurately to collate, summarize and store the information from the crime scene, including written records and preserved crime scene specimens, along with any you later preserved as the insect life cycle progressed.

A forensic entomology report may be requested for one of several purposes. These include: a report written in order that the evidence may be assessed to determine its strength and value, to assist solicitors and barristers; a report for individuals in response to an entomological problem which could, at a later date, be part of court proceedings in a civil court; the formal *Statement of Witness* which is submitted for consideration by the court during the actual proceedings in criminal and civil courts. In each of these instances, a logical pattern of explanation and clarity of information is needed.

The first two reports do not have a particular style which must be followed, although they must be accurate, objective and clearly written. The Statement of Witness has components which must be included, even though the style of the statement has not been laid down definitively. However, on all counts your responsibility is to the court. It is not to the 'party' who is instructing or paying you (Civil Procedure Rules 35 3.1. and 2), neither is it your role to assist in 'proving someone guilty'. There are suggestions which assist in presenting a Statement of Witness or report in an objective manner and they are described in the next section.

10.1   The Statement of Witness

The style of the Statement of Witness described below relates to the legal system which operates in England and Wales. The style of a report for the court may differ in other countries. At present there is no expected format for presenting forensic entomology statements of witness, although recently the Institute of Expert Witnesses and the Academy of Experts jointly proposed a generic format (2005), an 'Experts Protocol' which, despite it being intended for statements for the civil courts, is a basis for good practice. The guidance for this report style is the *Civil Procedure Rules, Part 35* and the accompanying *Practice Direction PD 35*. This 'Experts' Protocol' is in force from 5 September 2005. Coincidentally, in the European Union, a Code of Practice for Experts was brought in, which covers all of Europe.

The report structure has to follow a logical thread, so that the non-expert can understand it (Table 10.1). Under the relevant acts, the statement could potentially be read aloud in court, unless it is directed that the account can be oral, so the

Table 10.1   Suggested headings for a forensic entomology Statement of Witness

Component No.	Section topic in the statement	Page No.
1	*Title sheet* – reference number; your name; date; institution name and address; name of the court receiving the report; instigators of the report	Top right, Criminal Statement Centre bottom, Civil Statement
2	*Contents page* – indicate both paragraph number and page number	Sign each page
3	*Introduction* – who you are; age, usually 'over 18' for an expert witness; your qualifications	
4	*Summary of background* – the nature of the case and linking to where you sample	
5	*Instructions* – what you were asked to do; by whom. Do not stray from these instructions	

6	*Summary of conclusions* – very short sentences to direct the reader's understanding of what you found
7	*Description of what you sampled at the crime scene* – e.g. if it was a crime scene with a dead body, how you sampled for puparia; temperature assessments at the time; and specimens taken for additional analysis, such as bullet residues or DNA
8	*Investigations at the crime scene* – further meteorological analysis in the 3–5 days after finding the body
9	*Follow-up investigations or research*
10	*Meteorological data sources* or research if for a civil case
11	*Conclusions/opinion* – clear explanation; indicate the limitations and confirm with a signature. For civil cases include a Statement of Truth here
12	*Annex* – glossary; life cycles; sampling and larval culturing methodology; photographs and sketches; time line; CVs

report has to be written very clearly and simply. With the agreement of the judge this need not necessarily happen and expert witnesses can, instead, be called to court to explain the contents of their statement. This will be your experience when you role-play being an expert witness in a mock court.

For clarity, each statement or report should have a number of sections and should start with a front sheet which clearly specifies:

- *Status of the report*, i.e. is it for evidence assessment or a Statement of Witness?

- *To whom it is addressed* (reports for use as evidence, according to the Civil Procedure Rules Practice Direction (PD) 35 Para 2, should be *addressed to the court*, so indicate its name and location).

- *Who the report (or statement) is from* (your name should be in full) and the fact that you are a forensic entomologist. In addition, indicate if you are employed in any other capacity.

- *Date* for the day you sign and send out the report (sign all of the pages on that date).

- *Whether you are a joint single expert* or have been instructed as an expert witness by one *party to the case* or the other.

- The people who have issued *the instructions* to undertake the work, i.e. both the individual and the solicitors concerned or the name of the police force.

- Your *professional details*, i.e. name, professional address and contact details.

- *Court reference number*, if this is relevant to the report style.

This page should be followed by a *Contents* page. The next page may be a *Glossary* of terms in alphabetical order, or the glossary may be included with the Annex. A Glossary should be included so that the entomological terms can be looked up. It is useful to embolden in the text the technical words whose meaning can be found in the Glossary, as this helps to speed up reading the report.

The body of the forensic entomology report may be divided into 10 sections, the first of which is an *Introduction*:

1. *Introduction – who you are and why you can offer an opinion.* The purpose of this section is to indicate who you – the expert witness – are, your age and also your qualifications to express an opinion. In the case of a murder, the relevant acts, such as the Criminal Justice Act 1967 (Section 9), along with a Statement of Truth and a Statement of Compliance, can be presented as a preamble on the top of the page. This is because, in line with Section 22 of the Rules and Practice Direction and Rule 32.14, in the English courts, a *Statement of Truth* and a *Statement of Compliance* need to be included as a preamble if the Statement of Witness is for a criminal court (Section 9, Criminal Procedures Act 1967). (A Statement of Truth is placed at the end of the Statement of Witness, if the statement is for a civil court).

   The mandatory wording of the Statement of Truth is:

   > 'I confirm that insofar as the facts stated in my report are within my own knowledge I have made known which they are and I believe them to be true, and that the opinions I have expressed represent my true and complete professional opinion.'

   The format of the Statement of Compliance specified in PD 35 2.2(9) is as follows:

   > 'I understand that my duty is to the court and have complied and will continue to comply with that duty.'

   Underneath these statements put your signature and the date the report is sent out. Your qualifications and the names of anyone who provided technical or crime scene assistance should be included at this point.

The next heading in the Statement of Witness should be the summary of the background:

2. *Summary of the background* – the matter upon which you are asked to offer an opinion. This section is a short review of the relevant known facts of the case which relate to the question being asked. The police, or person who is instructing you, will provide these details.

3. *Instructions* – what you have been asked to do. In this section you indicate what you were asked to do and the issues of the case. You should not stray outside these boundaries in terms of what you actually do. Where you have suggested further analysis and are awaiting the results, you should submit an interim Statement of Witness and say that you would reconsider your conclusions in the light of the additional evidence.

This section is followed by:

4. *A summary of conclusions* that you have drawn.

5. *A description of what you sampled in response to the instructions*, e.g. you might indicate that you sampled orifices of the partially clothed body of a Caucasian female to recover insect larvae, or that you received photographs and specimens of 15 cockroaches from a food storage warehouse in Basingstoke.

6. *Investigations at the locality, or crime scene* – what and where you sampled. In this section you describe where you sampled if you attended the scene; the temperature recordings and soil surveying for any buried puparia; any other investigations which happened at the crime scene at the time you attended or subsequently. For example, if the samples were collected in the mortuary (morgue), then this is described in this section. Or if this is a civil case, any investigations you undertook at a factory, for example, should be included here.

7. *Follow-up investigations*. The content of this section describes any laboratory or additional investigations which you carried out, or which you requested to be undertaken by your client. Such investigations might include additional temperature recordings taken over 3–5 days at the crime scene after the body was recovered, or your requests that the larval gut contents be examined to identify the source of the DNA. The section may include details of your approach to determining the different lengths of the larvae from each sampling site, to demonstrate age and so which part of the body was initially colonized. The conditions used for culturing eggs or larvae to adulthood, or through to the life stage found on the body, should also be indicated in this section. Specifying the identity of the specimens provides a logical link to help you to demonstrate the duration of the post mortem

interval, using the estimated conditions at the crime scene before the body was discovered.

8. *Meteorological data sources.* Indicate in this section from where you sought the information about temperatures between when the body was found and the time the individual was last seen. For example, was it a local meteorological station or an amateur weather station that you used? How far away from the crime scene was the source of the weather details recorded? In a civil Statement of Witness, this section may not be relevant.

9. *Experimental analysis of relevant entomological data.* Indicate under this heading the references you have used to investigate the ecology of the insects infesting the body, food or domicile. For example, you may cite specifically the titles and authors of published papers you used to confirm the duration of life cycles of specific insects, and any taxonomy articles which helped you confirm the identity of the species of insect. The actual list of references should be provided in the Annex.

The next section is your deductions and conclusions about the time of death, or about the implication of the presence of the insects you have been investigating.

10. *Conclusions* – sometimes called *Opinion.* This section is where you link together the issues you were instructed to address, with the results you have recorded in the experimental and investigative sections of your report. It is more comprehensive than the summary you presented earlier. You should indicate the limits of your investigation, along with your estimate of the minimum time since death and an indication, where possible, of the level of accuracy of your findings.

11. *Annex* – material which helps explain additional aspects of the report. In this section you should include any supportive material. This might be a generalized life cycle of the specific fly species which you have identified, photographs, diagrams and maps of the crime scene and of the body, to indicate where the larvae were infesting the body, or of the investigations for a civil prosecution.

You may also choose to highlight pertinent sections from the academic references used to draw your conclusions and provide an explanation of what physiological, or thermal, energy is to explain ADD and ADH. Your actual calculations can be retained with your laboratory and crime scene records. They will be requested by the court if the need arises. So keeping clear, readable laboratory and crime scene notebooks is absolutely essential.

It is also useful to provide a time line of the entomological activities you have undertaken to reach your conclusion. This helps you show when the samples were collected, when they reached the laboratory, your sampling regime for the life stages and when the adults emerged, or when you started to breed through to the life stage of the insect collected from the body.

If you have been assisted in undertaking the work, the assistant's experience and qualifications (CV) should also be included in the Annex.

The report should be double-spaced on A4 paper (or foolscap in the USA) of a reasonable weight. The margins down each side of the page should be set at 3.5 cm, so that comments can be written in them. Your report should be word-processed on one side of each page, ideally using 14 point font so that the text can be quickly and easily read (Figure 10.1). Traditionally in English courts the font used is Times New Roman. Both the pages and the paragraphs should be numbered sequentially. Finally, the report should be checked and double-checked for accuracy of content and for spelling mistakes. Any inconsistencies and inaccuracies provide a target for questioning your work, or for necessitating your attendance at court. They will almost certainly be picked up in your mock court practice and make your experience all the more fraught!

For the sake of continuity and integrity of evidence it is good practice to sign and date each page (for a criminal case your signature should be witnessed), to prevent any question of pages being added or removed later. This should include those pages in the Annex. (Pages are numbered on the top right corner for criminal statements of witness and on the bottom, at the centre of the page, for civil reports and statements.) You should find out how many copies of the report are needed by the court and print out sufficient copies, ideally using a laser printer. This is better than photocopying them as it ensures all the reports look professionally presented.

The completed Statement of Witness (Figure 10.1) should be hole-punched so it can fit into a lever arch file or become part of the court bundle, and ideally it should be submitted as a bound copy between acetate sheets within a slip binder. This makes it easier to disassemble and to become part of the 'court bundle' retained in a ring binder. It also means you can handle the statement more easily if you need to refer to it in the 'witness box' and you drop pages less easily.

10.1.1   The Statement of Witness as a tool in the courtroom

The court may wish to ask the forensic entomologist to attend in person, to expand on the points in their Statement of Witness. This will certainly be the case in a mock courtroom role play. You will first be asked your full name and address and to state your qualifications, so be prepared to explain the meaning of the acronyms for the qualifications which you hold. The court may also be reassured to learn that you have been keeping up to date by regularly attending entomology and forensic entomology conferences, contributing to learned journals and also about any membership of professional societies which you hold. Next you may be required to clarify further any points in your report, although you will generally have been asked initially to summarize your findings and conclusions.

Statement of Witness

Statement of: Alwyn Campion   *Age* : Over 18

Occupation: FORENSIC ENTOMOLOGIST and CRIME SCENE ADVISOR

The Forensic Unit, University of East Lindsey, Brayford, Poolby, Lindseyshire LB4 6FT UK.

This statement consisting of 5 pages each signed by me, is true to the best of my knowledge and belief and I make it knowing that, if it is tendered in evidence, I shall be liable to prosecution if I wilfully stated in it anything which I know to be false or do not believe to be true.

(Criminal Justice Act 1967 section 9; Magistrates' Courts Act 1980 sub section 5A (3A) and 5B; Magistrates' Courts Rules 1981, Rule 70)

Signature: **Alwyn Campion**   Date: 30.6.2005

1. Introduction

I am a Principal Lecturer at the University of East Lindsey. I have a B.Sc. (Hons) Zoology, a PhD in insect ecology and am a Fellow of the Royal Entomological Society and a member of the Professional Association for Forensic Entomology. I have worked in Forensic Entomology for 15 years and have long experience of the use of Calliphoridae in determining the Post mortem interval. My research interest is the ecological interactions of flies and beetles with the corpse.

2. Background Summary

On May 29th at 11 am, I was asked by Lindsey Regional Police Consortium to attend a scene where the body of a Caucasian male had been discovered on the railway bank between Haswell and Brigg, West Lindsey. I was subsequently advised that the body had been identified as that of Jason Smith of Brigg, West Lindsey.

3. Instructions

I was asked to collect entomological samples from the body and meteorological data from the scene to estimate time since death.

4. Summary of Conclusions

i. Thirty six larvae were recovered from two sites on the body; 30 from beneath the eyelids (15 larvae were recovered for culturing to the adult stage and 15 larvae were preserved in Kahle's solution; and six from the wound on a hand. (Of these four larvae were retained for further culturing and two larvae were preserved in Kahle's solution). (see Annex 3 for preservation and culturing methodology)

ii. In the laboratory all the specimens were identified as second instar larvae of the Blowfly species *Calliphora vomitoria* (L).

iii. The estimated minimum time since death was 38 ± 8 hours. (See annex 2 for meterological data and annex 4 for background to PMI determination).

5. Description of what was sampled at the Crime Scene

Alwyn Campion   : 30.6.2005   Witnessed by : Jonathan Rogers

Figure 10.1   A specimen first page from a Statement of Witness for a criminal case

Many forensic practitioners in the UK also hold registration as expert witnesses through the Council for the Registration of Forensic Practitioners, as this can be seen as a measure of desire for professionalism.

10.2 Council for the Registration of Forensic Practitioners

This council (CRFP) was set up to allow self-regulation of forensic practitioners and to provide some degree of reassurance for the general populace, including the courts, about the practitioner's forensic professionalism. Currently, the specialisms of those registered include scenes of crime officers, document examiners and forensic toxicologists. These specialisms have been divided by the Council into three broad categories: Incident Investigation, Medicine and Healthcare, and Science and Engineering. The areas of practice are allocated within these categories. The council operates through voluntary peer review and requires that those seeking registration submit a portfolio of examples of their casework, as well as completing a self-examination questionnaire. The forensic expert also needs to have been regularly employed as an expert witness in the recent past. To become a registered forensic expert, the review requires you, the practitioner, to comply with a number of aspects or tenets which need to be upheld in your work and life style. These tenets are to:

1. Recognize that your overriding duty is to the court and to the administration of justice: it is your duty to present your findings and evidence, whether written or oral, in a fair and impartial manner.

2. Act with honesty, integrity, objectivity and impartiality.

3. Not discriminate on grounds of race, beliefs, gender, language, sexual orientation, social status, age, lifestyle or political persuasion.

4. Comply with the code of conduct of any professional body of which you are a member.

5. Provide expert advice and evidence only within the limits of your professional competence and only when fit to do so.

6. Inform a suitable person or authority, in confidence where appropriate, if you have good grounds for believing there is a situation which may result in a miscarriage of justice.

In all aspects of your work as a provider of forensic advice and evidence you must:

7. Take all reasonable steps to maintain and develop your professional competence, taking account of material research and developments within the relevant field and practising techniques of quality assurance.

8. Declare to your client, patient or employer, if you have one, any prior involvement or personal interest which gives, or may give, rise to a conflict of interest, real or perceived; and act in such a case only with their explicit written consent.

9. Take all reasonable steps to ensure access to all available evidential materials which are relevant to the examinations requested; to establish, so far as reasonably practicable, whether any may have been compromised before coming into your possession; and to ensure their integrity and security are maintained whilst in your possession.

10. Accept responsibility for all work done under your supervision, direct or indirect.

11. Conduct all work in accordance with the established principles of your profession, using methods of proven validity and appropriate equipment and materials.

12. Make and retain full, contemporaneous, clear and accurate records of the examinations you conduct, your methods and your results, in sufficient detail for another forensic practitioner competent in the same area of work to review your work independently.

13. Report clearly, comprehensively and impartially, setting out or stating:

    (a) Your terms of reference and the source of your instructions.

    (b) The material upon which you based your investigation and conclusions.

    (c) Summaries of your and your team's work, results and conclusions.

    (d) Any ways in which your investigations or conclusions were limited by external factors, especially if your access to relevant material was restricted; or if you believe unreasonable limitations on your time, or on the human, physical or financial resources available to you, have significantly compromised the quality of your work.

    (e) That you have carried out your work and prepared your report in accordance with this Code.

14. Reconsider and, if necessary, be prepared to change your conclusions, opinions or advice and to reinterpret your findings in the light of new information

or new developments in the relevant field; and take the initiative in informing your client or employer promptly of any such change.

15. Preserve confidentiality unless:

    (a) The client or patient explicitly authorizes you to disclose something.

    (b) A court or tribunal orders disclosure.

    (c) The law obliges disclosure.

    (d) Your overriding duty to the court and to the administration of justice demand disclosure.

16. Preserve legal professional privilege: only the client may waive this. It protects communications, oral and written, between professional legal advisers and their clients; and between those advisers and expert witnesses in connection with the giving of legal advice, or in connection with, or in contemplation of, legal proceedings and for the purposes of those proceedings.

CRFP Rules 1998

Cases requiring skills in forensic entomology are not commonplace and you may therefore have assisted at fewer cases than the CRFP prescribe. This may limit your eligibility to join the CRFP. However, the tenets of the council still provide good professional and ethical guidelines for forensic entomologists.

10.3 Communicating entomological facts in court

When you take part in your mock court training, or if you are asked to attend court, you will be expected to explain your findings in a way understandable to people from a wide variety of backgrounds. Most members of the general public appear frightened by insects, expecting them to sting or bite, so their general understanding of the life cycle of insects can be small. However, many crime-related television series have provided a useful background. Most people now know that maggots may be used to predict the time since death of a person and the discovery of their body.

It is necessary to routinely provide details of such things as life cycles for the particular species in question and the criteria for ageing maggots (larvae), as part of the Annex to your Statement of Witness. It is also helpful to have ready explanations for the meanings of the words used, or to use accurate but more familiar terminology in your explanations. Although as the expert witness in forensic entomology, you should not use jargon, sometimes the use of technical terms is unavoidable. The advice 'do not stray outside your expertise or be drawn to do this' is an important guide as you frame your responses, both in your mock court training and when acting as an expert 'for real'.

10.4 Physical evidence: its continuity and integrity

One of the crucial aspects of forensic science, and especially forensic entomology, is to ensure the continuity of the evidence. This is also true as the insect life stages are grown up in order to answer questions about the post mortem interval. The crime scene itself may provide information along with the insects inhabiting the body, so records, sketches, photographs and specimens should be taken and your sampling recorded. Some insect specimens only come to light after the body has been transported to the mortuary (morgue), either when the body bag is opened by other forensic scientists, or within the body on initial examination or dissection. This information, too, must be recorded if you, the forensic entomologist, are called to the mortuary or later receive further specimens relating to the case.

Some forensic entomologists attend the scene in person, take samples, culture the specimens gathered from the body and calculate the post mortem interval themselves, so the evidence never leaves their possession. In this instance there are fewer problems of continuity and integrity of evidence, provided that your actions and presence are logged both by you and in the scene investigating officer's log and that the two logs coincide. However, if crime scene investigators (SOCOs) collect the information, then it is imperative that they follow a recognized protocol, record their actions faithfully and carry out the appropriate packaging procedures, separately sealing the live and preserved specimens into containers for each location, from each collecting site on the body. These crime scene investigators (or members of the police) must also ensure the integrity of the preserved samples and live cultures, keeping the cultures alive but below their base temperature, until the samples, along with the relevant crime scene notes, reach the forensic entomologist.

If the entomological samples have been collected by the crime scene investigator, pathologist or another forensic scientist, it is important that that information recorded about the contents of the packages of samples is very accurate. Samples of preserved larvae, together with the live specimens from that site on the body, should be packaged together and identified with an individual identity number (or, for example, a bar code) that is separate from those collected from a different location on the body.

On transporting and receiving the packages, the same records of contents, numbers and locations must be made to ensure the integrity of the information. For example, to describe the contents of a package as 'three sample pots, each containing live larvae with liver plus three sample pots of preserved larval specimens collected from the corpse of L.T. Haslen' leaves less room for challenge than describing the content of the package as 'six containers of maggots collected from the corpse of L.T. Haslen'.

This integrity of samples and the notes about their collection is also important. In every instance, the specimens and sites must be photographed and these notes and photographs, and those made subsequently in the laboratory, must all be catalogued. From this information a report, or Statement of Witness, can be generated which will be considered accurate and comprehensive. Your laboratory notebook provides support material to demonstrate the continuity and integrity of the evidence if a question arises.

A further point to consider when presenting the report in court is the nature of the illustrations which you have chosen to use. Pictures, whilst illustrating where the samples were taken from the body, may also be quite disturbing if shown to members of the general public present in court. When you assemble the evidence to present to the mock court (or actual court), it is necessary to choose carefully the particular photographs which best illustrate the scientific point you wish to make. It may even be helpful to give verbal warnings about the content of any photographs prior to showing them to the solicitor or barrister, as well as in court. As Greenberg and Kunich (2002) also point out, in describing the views of a Supreme Court judge in Hawaii, the forensic entomologist should be aware that for those less familiar with corpse consumption by maggots, such pictures may influence their feelings about the individual accused of the crime. The use of black and white photographs may be less prejudicial.

The approach of the forensic entomologist in court is best summarized by the second tenet proposed by the CRFP, which is: *'Act with honesty, integrity, objectivity and impartiality'*.

10.5 Review technique: writing a Statement of Witness using the post mortem calculations determined from details given in Chapter 7.

You have been instructed by the police to prepare a Statement of Witness for the case of the body of a young female found in the Pleasure Gardens, Wingsea, for which you estimated the post mortem interval (Chapter 7). The trial for murder (Regina vs Morgan) will be heard on 27 November. The suspect is a person called Clay Morgan, also of Wingsea, who was arrested in May. You are asked to provide a Section 9 Statement of Witness, giving an estimate of the post mortem interval. For the purposes of this practical task, it has been agreed with the police that photographic illustrations for the report will be submitted separately, since they were taken by the police photographer as you were collecting the samples, and are not included in your report. Use the format described in Chapter 10 to prepare your statement.

10.6 Further reading

Adair T. W. 1999. Chain of custody in the university setting: considerations for entomologists. *Antenna* **23**(3): 140–143.

Bartle R. 2002. *Police Witness: A Guide to Presenting Evidence in Court*. The New Police Bookshop, Surrey: www.policebooks.org.uk

Bond C., Solon M. and Harper P. 1999. *The Expert Witness in Court: A Practical Guide*, 2nd edn. Shaw & Sons: Crayford, Kent; pp. 74–126.

Civil Justice Council. 2005. *Protocol for the Instruction of Experts to Give Evidence in Court*. Institute of Expert Witnesses and Academy of Experts: London.

Greenberg B. and Kunich J. C. 2002. *Entomology and the Law: Flies as Forensic Indicators*. Cambridge University Press: Cambridge; pp 249–283.

Website addresses

The Academy of Experts: www.academy.experts.org.uk

Expert Witness Institute: www.ewi.org.uk/

Civil procedure rules, practice directions and protocols: http://www.dca.gov.uk / civil / prorules_fin / index.htm

Council for the Registration of Forensic Practitioners: www.crfp.org.uk

11

The role of professional associations for forensic entomologists

There are several professional associations for forensic entomologists and they are all designed to achieve two things. The first is to maintain a forum for discussion and debate to ensure that the professionalism of the forensic entomologist is maintained, not least through the exchange of information and the development and maintenance of best practice. The second role is one of maintenance of standards in practice.

Whilst the forensic entomologist may be a professional entomologist, there are no formal associations of the same regulatory nature as the UK Council for the Registration of Forensic Practitioners, or for accreditation of the forensic entomology practitioner. To date there is no group of forensic entomologists involved as a working group within the CRFP in the UK to develop standard protocols which are recognized, such as there is for Document Examiners, or Scenes of Crime Officers. Indeed, there is a self-selected list of individuals involved in forensic entomology; the Worldwide Directory of Forensic Entomologists, which currently lists under 100 members (www.missouri.edu~agwww/entomology/chapter1.html), but there is no indication of level of expertise, or court training, of these experts.

The two major associations which are seeking to move the development of standards forward are the American Board of Forensic Entomology and the European Association for Forensic Entomology.

11.1 Professional organizations

11.1.1 American Board of Forensic Entomology

This organization is a society which requires its members to have a postgraduate qualification and appropriate experience of forensic entomology and the crime scene. It developed as a result of a group of entomologists and parasitologists meeting together at an Entomological Society of America meeting, until eventually in 1996 the American Board of Forensic Entomology was formed (Goff, 2000). However, those practising forensic entomology may also include scientists who are members of the Entomological Society of America, the American Academy of Forensic Science and the American Registry of Professional Entomologists. No specific organization is the accredited association for forensic entomology in America.

11.1.2   European Association for Forensic Entomology (EAFE)

The European Association for Forensic Entomology was founded in 2002. The society was launched in France, which is apposite, since Mégnin was a Frenchman and his work formed part of the foundations of forensic entomology. The association has a number of aims:

- To seek a common protocol.

- To foster high standards of competency in specimen collection and analysis.

- To create a solid scientific basis so that forensic entomology can be a valid analytical tool.

In setting itself these goals, EAFE will encourage the recognition of forensic entomology as an appropriate crime scene investigative tool, which has a role in assisting, where relevant, the investigation of matters relating to the crime scene.

The association is young, with a burgeoning membership and serious intentions to achieve professional standards. It started on this route quite recently, despite forensic entomology having a foundation of at least 100 years of research. However, it must be recognized that only in 2004 did the Forensic Science Society, a UK organization which started in 1959 and is one of the oldest associations of forensic practitioners, actually choose to seek recognition as an accrediting body for forensic scientists.

11.2   Forensic entomology protocols

At present, because there is no mechanism for accreditation, standard protocols for forensic entomology are not utilized worldwide, although the protocols generated by Catts and Haskell (1990), and in a similar format by Byrd and Castner (2001), do form the general guide to working practice and provide a case study form to be completed at the crime scene. The European Association for Forensic Entomology has consistently highlighted the need to develop a sampling protocol for general agreement by forensic entomologists, and subsequent use by those crime scene investigators who collect samples at the crime scene on their behalf. The annual meeting of EAFE in Lausanne in 2005 went a great way to achieve this aim.

Workshops and field visits were provided from which sampling from a number of dead pigs and subsequent identification of the species of insects present was discussed. The delivery of these workshops, as part of the meeting, provided a basis whereby commonality of methods could be achieved and for exchange of approaches to good working practice. These protocols will be debated, tested and gradually become accepted by the community of forensic entomologists.

11.3 Areas for future research

Clearly, because crime scene investigation is context-dependent, forensic entomologists will define their own research requirements for their own localities. However, students (sixth formers, college and undergraduate) provide a rich supply of enthusiasts who can contribute to forensic entomology as part of their studies or degree programme. It is from amongst such students that the new generation of forensic entomologists will arise.

The areas for projects are varied and come both from a study of topics in ecological entomology and from investigating aspects of simulated crime scenes. There will be variation in the species which initially colonize a body at different locations, or which are part of the insect succession in different geographic areas or in different habitats – upland grasslands compared to meadows, bodies which are hanging or lay on or under the ground, for example. Archer (2004) provided a baseline for studies on carcasses on the ground surface in Australia, but more studies are needed in other places. Collecting local information about individual species of insects which colonize carrion, or even using a human equivalent such as a pig, by undertaking sampling in a number of areas in each country, is useful. Some of this work is achieved simply by trapping the species visiting carrion, but contributions to forensic entomology can also be made by investigating the distribution of a particular species in a country. Currently, biological records centres provide a basis for coordinating this information, building on databases of the distribution of insect species.

Further research to determine basal temperatures for carrion-visiting species is needed from different countries. This is true for both the more common species of insects found on the cadaver, such as *Calliphora vicina*, *Calliphora vomitoria* and *Lucilia sericata*, and also those species which are less frequently quoted, such as *Piophila casei* or *Lucilia caesar*. Such research ensures the rigour of the post mortem interval determination and reduces the chances of the time since death estimates being challenged.

Developments in DNA technology have fostered an interest in undertaking a genetic analysis of the larval gut content, illuminating the last meal of the larvae collected from a corpse. This can link a particular larval stage to a corpse. It can also be of great importance where other food sources are locally available and it is important to determine whether the presence of the insect species is linked to a scene of crime or the presence of a person in a natural disaster. DNA analysis of insect specimens can, under some circumstances, also be used where a chain of custody of some evidence is brought into question.

DNA analysis of gut contents can provide the basis of other work, and Campobasso *et al.* (2005) have reviewed the current position on this aspect, including the protocols for identifying insects from their DNA. For example, less work has been carried out on those arthropods which colonize submerged bodies and this technology might be utilized in the context of determining time since submergence of bodies.

Study of the dynamics of insects colonizing the corpse is another area which would benefit from further study. The influence of one insect species on the presence and survival of the rest can affect succession. A basis for this study has been provided by Hanski (1987), who explored carrion fly dynamics over a 4 year period in southern Finland. He investigated where on the body three families of initial colonizers were found and their local interactions. He confirmed that sarcophagids were poor competitors. The distribution of sarcophagid larvae initially on a corpse needs further clarification, since the first instar larvae are deposited one by one on the body and not when eggs hatch.

The effects of larval competition on entry into diapause is a valuable area, including whether one particular generation of larvae enters diapause more readily than another, since this has relevance to interpreting the insect life cycle found on a corpse. Haski suggested that setting up field experiments to investigate the influence of larval size in relation to diapause, on the corpses of a particular species using different-sized corpses, would be a productive area for this study. It has forensic applications, e.g. in terms of the influences of corpse size when colonizing juveniles rather than adult humans. Dr Zak Erzinçlioğlu, in his book entitled *Blowflies* (1996), highlights further areas of study relating to more ecological aspects of the behaviour of blowflies.

The ecology and carrion dynamics of the corpse and the relationship between species and habitats provide a source of further information from which to refine forensic deductions from forensic evidence in the field. The requirements for such studies are simple and the investigations can be low cost, whilst at the same time contributing greatly to the advancement of forensic entomology.

11.4 Further reading

Archer M. S. 2004. Annual variation in arrival and departure times of carrion insects at carcasses: implications for succession studies in forensic entomology. *Australian Journal of Zoology* **51**(6): 569–576.

Amendt J., Campobasso C. P., Gaudry E., Reiter C. *et al.* 2006. Best practice in forensic entomology – standards and guidelines. *International Journal of Legal Medicine* (in press).

Campobasso C. P., Linville J. G., Wells J. D. and Introna F. M. D. 2005. Forensic genetic analysis of insect gut contents. *American Journal of Forensic Medicine and Pathology* **26**(2): 161–165.

Disney R. H. L. 2005. Duration of development of two species of carrion-breeding scuttle flies and forensic implications. *Medical and Veterinary Entomology* **19**(2): 229–235.

Drake C. M. 1991. *Provisional Atlas of the Larger Brachycera (Diptera) of Britain and Ireland.* Biological Records Centre: Huntingdon.

Erzinçlioğlu Y. Z. 1996. *Blowflies.* Naturalists' Handbooks No. 23. Richmond Publishing: Slough.

Gibbs J. P. and Stanton E. J. 2001. Habitat fragmentation and arthropod community change: carrion beetles, phoretic mites and flies. *Ecological Applications* **11**(1): 79–85.

Hanski I. 1987. Carrion fly community dynamics: patchiness, seasonality and coexistence. *Ecological Entomology* **12**: 257–266.

Tabor K. L., Brewster C. C. and Fell R. D. 2004. Analysis of the successional patterns of insects on carrion in south-west Virginia. *Journal of Medical Entomology* **41**(4): 785–791.

Tobin P. C. and Bjørnstad O. N. 2003. Spatial dynamics and cross correlation in a transient predator–prey system. *Journal of Animal Ecology* **72**(3): 460–467.

Website sources

European Association for Forensic Entomology: www.eafe.org
American Board of Forensic Entomology: www.research.missouri.edu/entomology/

Appendices

Appendix 1: Form for forensic entomology questions to be asked at the crime scene

---

FORENSIC ENTOMOLOGY QUESTIONS TO BE ASKED AT THE CRIME SCENE

Location of the crime scene Date and time body found Name of victim if known _____ **Date and time last seen** _____

Who is the collector of the specimens? _____
Date of collection _____
Who is the officer in charge, or the person instructing you to investigate the scene? _____
What is the scene of death like, i.e. is it rural or urban; if inside, are any windows open or closed; if outside, in shade or full sunlight, what is the vegetation like, is the body buried or on the soil surface? (photographs are valuable)

_____
_____
_____
_____

What is the position of the body (sketch)?

Is the body clothed? If so, describe the nature and condition of the clothing?

What is the state of decomposition?

What is the temperature: on the body surface ___; 0.31 m above the body surface ___; 1.1 m above the body surface ___; beneath the body ___; soil temperature 10 cm _____ and 20 cm below the surface _____

Sketch diagram of the location of the samples taken from each infestation site on the body

Time	Sample No.	Description of sample

Comments

Appendix 2: Answers to the calculation of the post mortem interval for the body at the Pleasure Gardens, Wingsea

The temperatures in the data are low, so the experimental life cycle estimates on experimental growth of *Calliphora vomitoria* at 12.5°C (Greenberg and Kunich, 2002) are appropriate to use. From their figures, the data for *Calliphora vomitoria* are:

> Egg stage   64.8 hours minimum duration
> LI stage   55.2 hours minimum duration
> L2 stage   60.0 hours minimum duration

We do not know how long the larvae have been in the second stage, so the minimum duration will be to at least the end of the first instar if the second stage had newly emerged. The egg stage (64.8 hours) and the first larval instar minimum duration (55.2 hours) estimates were therefore used in the calculation, since this would ensure that the conclusions were based on demonstrable fact, i.e. age of larva.

> Total duration in hours (64.8 + 55.2)   120 hours
> Experimental constant temperature used   12.5°C
> Base temperature chosen   3°C

$$\text{ADH} = \text{time}_{\text{hours}} \times (\text{temperature} - \text{base temperature})$$

$$\text{ADH} = 120 \times (12.5 - 3)$$
$$= 120 \times 9.5$$
$$= 1140°\text{H}$$

From the information given in the sidebar on post mortem interval calculation, in Chapter 7, we can count back, using the sum of the actual accumulated degree hours which accrued, to estimate the minimum time since death. Each slot represents the amount of physiological energy generated per hour (ADH) added together. So by counting the slots we find that the estimated post mortem interval is 5 days 22 hours, because the ΣADH we are looking for is a value of 1140. The table values of relevance in the final column will be 1138.1 and 1151.9, i.e. the values relating to 142 hours since 11 am on 20 April, when the body was found. The figure of 1140 is more than the value 1138.1 but under the value of 1151.9 for the next hour slot. So the estimated post mortem interval is a minimum of 5 days 22 hours and the time of death would be at or after 2–3 p.m. on 14 April.

Appendix 3: UK list of Calliphoridae (2006)

Bellardia bayeri (Jacentkowský, 1937)
Bellardia pandia (Walker, 1849)
Onesia biseta (Kramer, 1917)
Bellardia pubicornis (Zetterstedt, [1838])   Scarce
Pseudonesia puberula (Zetterstedt, [1838])
Bellardia viarum (Robineau-Desvoidy, 1830)
Onesia pusilla (Meigen, 1826)
Bellardia vulgaris (Robineau-Desvoidy, 1830)
Onesia agilis (Meigen, 1826)
Calliphora loewi Enderlein, 1903   Scarce
Acrophaga alpine authors, misident.
Calliphora stelviana (Brauer and von Bergenstamm, 1891)
Calliphora subalpina (Ringdahl, 1931)
Calliphora uralensis Villeneuve, 1922
Calliphora vicina Robineau-Desvoidy, 1830
Calliphora vomitoria (Linnaeus, 1758)
Cynomya mortuorum (Linnaeus, 1761)
Phormia regina (Meigen, 1826)
Protocalliphora azurea (Fallén, 1817)
Protocalliphora sordida authors, misident.
Phormia terrae-novae (Robineau-Desvoidy, 1830)
Protophormia terraenovae (Robineau-Desvoidy, 1830)
Eurychaeta palpalis (Robineau-Desvoidy, 1830)
Helicobosca distinguenda (Villeneuve, 1924)
Lucilia ampullacea Villeneuve, 1922
Lucilia bufonivora Moniez, 1876
Lucilia caesar (Linnaeus, 1758)
Lucilia illustris (Meigen, 1826)
Lucilia richardsi Collin, in Richards, 1926
Lucilia sericata (Meigen, 1826)
Lucilia silvarum (Meigen, 1826)
Angioneura acerba (Meigen, 1838)
Angioneura cyrtoneurina (Zetterstedt, 1859)
Eggisops pecchiolii Rondani, 1862   Scarce
Melanomya nana (Meigen, 1826)
Morinia nana (Meigen, 1826)
Melinda anthracina Wainwright, 1928
Melinda gentiles Robineau-Desvoidy, 1830
Melinda caerulea Wainwright, 1928
Melinda viridicyanea (Robineau-Desvoidy, 1830)
Pollenia amentaria (Scopoli, 1763)

Pollenia vespillo authors, misident.
Pollenia angustigena Wainwright, 1940
Pollenia griseotomentosa (Jacentkovský, 1944)
Pollenia varia authors, misident.
Pollenia excarinata Wainwright, 1940
Pollenia labialis Robineau-Desvoidy, 1863
Pollenia pediculata Macquart, 1834
Pollenia rudis (Fabricius, 1794)
Pollenia vagabunda (Meigen, 1826)
Pollenia carinata Wainwright, 1940
Pollenia viatica Robineau-Desvoidy, 1830
Stomorhina lunata (Fabricius, 1805)

Source: www.mapmate.co.uk [accessed 15:54 Mon 09 Jan 2006]. Reproduced by permission of Teknica Ltd

Appendix 4: UK checklists for Coleoptera

Not all of the beetles listed are carrion feeders; however, the whole list is reproduced for the sake of completeness.

UK species checklist for Trogidae (from www.mapmate.co.uk: reproduced by permission of Teknica Ltd)

Trox perlatus Goeze, 1777
Trox sabulosus (Linnaeus, 1758)
Trox scaber (Linnaeus, 1767)

UK species checklist for Histeridae (from www.mapmate.co.uk: reproduced by permission of Teknica Ltd)

Abraeus globosus (Hoffmann, J., 1803)
Abraeus granulum Erichson, 1839
Abraeus perpusillus (Marsham, 1802)
Acritus homoeopathicus Wollaston, 1857
Acritus nigricornis (Hoffmann, J., 1803)
Aeletes atomarius (Aubé, 1842)
Atholus bimaculatus (Linnaeus, 1758)
Atholus duodecimstriatus (Schrank, 1781)
Carcinops pumilio (Erichson, 1834)
Dendrophilus punctatus (Herbst, 1792)
Dendrophilus pygmaeus (Linnaeus, 1758)
Dendrophilus xavieri Marseul, 1873
Epierus comptus Erichson, 1834
Gnathoncus buyssoni Auzat, 1917
Gnathoncus communis (Marseul, 1862)
Gnathoncus nannetensis (Marseul, 1862)
Gnathoncus nanus (Scriba, 1790) non Piller and Mitterpacher, 1783
Gnathoncus rotundatus (Kugelann, 1792)
Gnathoncus schmidti Reitter, 1894
Halacritus punctum (Aubé, 1842)
Hetaerius ferrugineus (Olivier, 1789)
Hister bissexstriatus Fabricius, 1801
Hister illigeri Duftschmid, 1805
Hister impressus Fabricius, 1798
Hister quadrimaculatus Linnaeus, 1758
Hister quadrinotatus Scriba, 1790
Hister unicolor Linnaeus, 1758
Hypocaccus dimidiatus (Illiger, 1807)
Hypocaccus metallicus (Herbst, 1792)

Hypocaccus rugiceps (Duftschmid, 1805)
Hypocaccus rugifrons (Paykull, 1798)
Kissister minimus (Aubé, 1850)
Margarinotus brunneus (Fabricius, 1775)
Margarinotus marginatus (Erichson, 1834)
Margarinotus merdarius (Hoffmann, J., 1803)
Margarinotus neglectus (Germar, 1813)
Margarinotus obscurus (Kugelann, 1792)
Margarinotus purpurascens (Herbst, 1792)
Margarinotus striola (Sahlberg, C.R., 1819)
Margarinotus ventralis (Marseul, 1854)
Myrmetes paykulli Kanaar, 1979
Myrmetes piceus (Paykull, 1809) non Marsham, 1802
Onthophilus punctatus (Müller, O.F., 1776)
Onthophilus striatus (Forster, 1771)
Paralister carbonarius sensu auctt. Brit. non Hoffmann, J., 1803
Paromalus flavicornis (Herbst, 1792)
Paromalus parallelepipedus (Herbst, 1792)
Plegaderus dissectus Erichson, 1839
Plegaderus vulneratus (Panzer, 1796)
Saprinus aeneus (Fabricius, 1775)
Saprinus cuspidatus Ihssen, 1949
Saprinus immundus (Gyllenhal, 1827)
Saprinus planiusculus Motschulsky, 1849
Saprinus semistriatus (Scriba, 1790)
Saprinus subnitescens Bickhardt, 1909
Saprinus virescens (Paykull, 1798)
Teretrius fabricii Mazur, 1972

UK checklist for Silphidae (from www.mapmate.co.uk: reproduced by permission of Teknica Ltd)

Aclypea opaca (Linnaeus, 1758)
Aclypea undata (Müller, O.F., 1776)
Dendroxena quadrimaculata (Scopoli, 1772)
Necrodes littoralis (Linnaeus, 1758)
Nicrophorus germanicus (Linnaeus, 1758)
Nicrophorus humator (Gleditsch, 1767)
Nicrophorus interruptus Stephens, 1830
Nicrophorus investigator Zetterstedt, 1824
Nicrophorus vespillo (Linnaeus, 1758)
Nicrophorus vespilloides Herbst, 1783
Nicrophorus vestigator Herschel, 1807

Oiceoptoma thoracicum (Linnaeus, 1758)
Silpha atrata Linnaeus, 1758
Silpha carinata Herbst, 1783
Silpha laevigata Fabricius, 1775
Silpha obscura Linnaeus, 1758
Silpha tristis Illiger, 1798
Silpha tyrolensis Laicharting, 1781
Thanatophilus dispar (Herbst, 1793)
Thanatophilus rugosus (Linnaeus, 1758)
Thanatophilus sinuatus (Fabricius, 1775)

UK Checklist of Dermestidae (from www.mapmate.co.uk: reproduced by permission of Teknica Ltd)

Anthrenocerus australis (Hope, 1843)	Australian carpet beetle
Anthrenus coloratus Reitter, 1881	
Anthrenus flavipes LeConte, 1854	
Anthrenus fuscus Olivier, 1789	
Anthrenus museorum (Linnaeus, 1761)	Museum beetle
Anthrenus olgae Kalik, 1946	
Anthrenus pimpinellae (Fabricius, 1775)	
Anthrenus sarnicus Mroczkowski, 1962	
Anthrenus scrophulariae (Linnaeus, 1758)	
Anthrenus verbasci (Linnaeus, 1767)	
Attagenus brunneus Faldermann, 1835	
Attagenus cyphonoides Reitter, 1881	
Attagenus fasciatus (Thunberg, 1795)	
Attagenus pellio (Linnaeus, 1758)	Two-spotted carpet beetle
Attagenus smirnovi Zhantiev, 1973	
Attagenus trifasciatus (Fabricius, 1787)	
Attagenus unicolor (Brahm, 1791)	Black carpet beetle
Ctesias serra (Fabricius, 1792)	Cobweb beetle
Dermestes ater De Geer, 1774	
Dermestes carnivorus Fabricius, 1775	
Dermestes frischii Kugelann, 1792	
Dermestes haemorrhoidalis Küster, 1852	
Dermestes lardarius Linnaeus, 1758	Bacon beetle
Dermestes leechi Kalik, 1952	
Dermestes maculatus De Geer, 1774	Hide beetle
Dermestes murinus Linnaeus, 1758	
Dermestes peruvianus Laporte de Castelnau, 1840	
Dermestes undulatus Brahm, 1790	

Globicornis nigripes (Fabricius, 1792)
Megatoma undata (Linnaeus, 1758)
Orphinus fulvipes (Guérin-Méneville, 1838)
Reesa vespulae (Milliron, 1939)
Thorictodes heydeni Reitter, 1875
Thylodrias contractus Motschulsky, 1839
Trinodes hirtus (Fabricius, 1781)
Trogoderma angustum (Solier, 1849)
Trogoderma glabrum (Herbst, 1783)
Trogoderma granarium Everts, 1898
Trogoderma inclusum LeConte, 1854   Larger cabinet beetle
Trogoderma variabile Ballion, 1878

UK checklist of Cleridae (from www.mapmate.co.uk: reproduced by permission
of Teknica Ltd)

Korynetes analis Klug, 1840
Korynetes caeruleus (De Geer, 1775)
Necrobia ruficollis (Fabricius, 1775)
Necrobia rufipes (De Geer, 1775)
Necrobia violacea (Linnaeus, 1758)
Opilo mollis (Linnaeus, 1758)
Paratillus carus (Newman, 1840)
Tarsostenus univittatus (Rossi, 1792)
Thanasimus formicarius (Linnaeus, 1758)   Ant beetle
Thanasimus rufipes (Brahm, 1797)
Thaneroclerus buquet (Lefebvre, 1835)
Tilloidea unifasciatus (Fabricius, 1787)
Tillus elongatus (Linnaeus, 1758)
Trichodes alvearius (Fabricius, 1792)
Trichodes apiarius (Linnaeus, 1758)   Bee-eating beetle

Appendix 5: List of relevant UK legal acts and orders

Anatomy Act 1984

The original Anatomy Act came into being because of grave robbing. The 1984 Act detailed the laws and, along with the Anatomy Regulations 1988, provided the procedures for:

- Allowing a person to have a licence to carry out anatomical examinations until aged 70.

- Licencing a premises to be used for carrying out anatomical examination.

- Providing permission for a body to be used for anatomical examination.

Human Tissue Act 2004

The law relating to the Human Tissue Act 1961, the Anatomy Act 1984 and the Human Organ Transplants Act 1989 was revisited recently. The current Human Tissues Act 2004 was enacted to unify the procedures relating to human tissue and it replaces the above acts. Schedule Part 3 Clause 45 provides for the disposal of human tissue material – decent disposal.

Animal By-products Regulations 2003

In the UK these are administered by the Department for Environment, Food and Rural Affairs (DEFRA). These regulations must be considered, and responded to, if forensic entomology experiments are to be undertaken. Discussion with the local DEFRA offices will usually provide information and support, and also guidance on organizing experiments under 'General Orders'.

Water Resources Act 1991

These provide waste water disposal regulations. The Act dictates what can be disposed of down storm drains and what has to be disposed of by means of the foul sewers (in Universities, the Estates Department will have a diagram of the locations of the foul sewers for the relevant properties). Waste water disposal cannot be directly into ditches, where there is the possibility that contaminated water can seep down to the water table and from there into our water resources.

Practical aspects of the acts and regulations for higher education

A number of aspects of the acts and regulations cited above relate to how university practicals have to be set up in order to conform to the requirements:

- Source of animals, e.g. pigs as representative human models.

- Fallen stock has to be disposed of by burning in an incinerator or by rendering. This has implications about the source of any model humans which might be used.

- The Welfare of Animals (Slaughter and Killing Regulations) 1955 means that animals awaiting slaughter should not be subjected to pain or suffering and should not be caused 'avoidable excitement'.

In summary, the relevant regulations mean:

1. Attendance at human post mortems is a privilege and they may be difficult to arrange for forensic entomology students to attend a post mortem in person.

2. A pig may be used as a model human but the source of the animal is strictly controlled. As indicated in the earlier paragraph, fallen stock has to be disposed of by burning it. This has implications for the end of experiments.

3. Any such model pigs must be kept under a cage, since dispersal as a result of scavengers such as birds and rodents is illegal. The parts of the body should be contained away from the sides of the cage. The wire mesh must be big enough to permit access by insects but prevent larger creatures gaining access.

4. Disposal of the carcass at the end of the study activities must be via a licensed carrier who can burn the carcass, which has to be stored in suitable containers en route.

5. Only those licensed can move the animal by-products around. This means that the body has to be collected from the research site.

Cages should be washed in a disinfectant that is recognized by DEFRA to control those diseases which can be transmitted from carcasses (DEFRA website has access to a list). A substance such as double-strength Milton is acceptable, but the items require steeping for a set period of time for satisfactory disinfection.

The washing water has to be disposed of down the foul sewer. If Milton is used, then there is no particular problem in disposing of the usual amount required for washing a cage.

Appendix 6: Selected sources of entomological equipment

(Further addresses may be found on the web)

Alana Ecology Ltd
The Old Primary School, Church Street, Bishop's Castle, Shropshire SY9 5AE, UK
Tel: +44 (0)1588 630173
Fax: +44 (0)1588 630176
www.alanaecology.com/
E-mail: sales@alanaecology.com

Arnold Johnson (for insect cages)
1 Bron-y-Glyn, Bronwydd, Carmarthen SA33 6JB, Dyfedd, UK
Tel: +44 (0) 1267 236329

BioQuip
17803 LaSalle Avenue, Gardena, CA 90248-3602, USA
Tel: 310-324-0620
Fax: 310-324-7931
www.bioquip.com/
E-mail: bioquip@aol.com

B and S Entomological Services
37 Derrycarne Road, Portadown, Co. Armagh, Northern Ireland BT62 1PT, UK
Tel: +44 (0)77 6738 6751 or +44 (0)28 3833 6922
Fax: +44 (0)28 3833 6922
www.entomology.org.uk/
E-mail: enquiries@entomology.org.uk

Carolina Biological Supply
2700 York Road, Burlington, NC 27215, USA
Tel: 800-334-5551
Fax: 919-584-3399

Entomorpho
Porfilio Giuseppe, Parma, Italy
Tel: +39 (0) 521 774720
Fax: +39 (0) 521 774720-Parma (Home)
E-mail: elitra@entomorpho.com

Henshaw
D. J. and D. Henshaw, 34 Rounton Road, Waltham Abbey, Essex EN 9 3AR, UK
Tel: +44 (0) 1992 717663
Fax: +44 (0) 1992 717663
E-mail: djhagro@aol.com

Watkins and Doncaster
The Naturalists, Robin J. Ford, P.O. Box 5, Cranbrook, Kent TN18 5EZ, UK
Tel: +44 (0) 580 753133
Fax: +44 (0) 580 754054
www.watdon.com/

Appendix 7: Legal information relevant to giving testimony as a forensic entomologist in the USA

Forensic entomologists provide expert testimony to courts as to the information which can be concluded from entomological (or related arthropod) evidence. In the USA expert evidence must meet a number of criteria as a result of a legal case, *Daubert-v-Merrell Dow Pharmaceuticals* Inc (1992) 509 US 579 test, where the criteria which must be applied to expert evidence were defined. An e-wire article, *Taking Experts to Court* (from www.jspubs.com [accessed 22 June 2005]: reproduced by permission of Dr C. Pamplin, UK Register of Expert Witnesses) summarizes the information surrounding decisions about expert witness testimony and is cited below:

> 'Rule 702 of the Federal Rules of Evidence (FRE) states that if scientific, technical or other specialized knowledge will assist the trier of fact to understand the evidence or to determine a fact in issue, a witness qualified as an expert by knowledge, skill, experience, training or education may testify thereto in the form of an opinion or otherwise. Difficulty arose over how this rule should be interpreted and applied. In Daubert, the court set out four criteria for determining whether expert testimony met the requirement that it constitutes "scientific knowledge". These are:
>
> - Whether the theory of technique "can be (and has been) tested".
>
> - Whether the "theory or technique has been subjected to peer review and publication".
>
> - In the case of a particular technique, what "the known or potential rate of error" is or has been.
>
> - Whether the evidence has gained widespread acceptance within the scientific community.
>
> In addition, the court may have to weigh the probative value of the evidence against its possible prejudice under Rule 403 FRE. This rule gives the judge a discretion to exclude expert testimony where its "probative value" is substantially outweighed by the "danger of unfair prejudice, confusion of the issues, or misleading the jury".'

Forensic entomology is able to meet these criteria on all counts and has been recognized as a field providing appropriate scientific evidence (Greenberg and Kunich, 2002).

Glossary

Adecticous – insects (or arthropods) which have immoveable (non-articulated) mouthparts. This is frequently true of the pupae and the mandibles may be reduced. Two types are known: those with appendages which do not attach to the rest of the pupal body, as in most of the Brachycera (called exarate adecticous pupae), and those where the appendages are firmly stuck to the pupal body, as is found in the Nematocera (called obtect adecticous pupae).

Adephaga – a suborder of Coleoptera (beetles) where the hind coxae are fixed to the metasternum and the first abdominal sternite is completely divided by the hind coxae.

Adipocere – grave wax, found on bodies in the first weeks or months after death because decomposition is taking place under damp conditions.

Arista – bristle-like projection from the third antennal segment. It can be feathery or with long or short hairs (pubescent).

Antenna – a structure on the head which detects odours. It is made up of a number of segments. The first two have names: the scape is the first antennal segment; the pedicel is the second antennal segment.

Appendages – a limb, e.g. a leg or wing, which is attached to the body by a joint.

Archostemata – a suborder in the classification of the Coleoptera, in which the larvae are mainly wood feeders.

Acrostichal – hairs found between the rows of dorso-central bristles on the top of the thorax.

Anaerobic – needs the absence of (or is not dependent upon the presence of) oxygen.

Bacteria – enucleate (prokaryotic) microorganisms, which in this instance are involved in decomposition.

Basicosta – the second epaulette or plate-like structure on the lead vein of the membranous wing. In *Calliphora* species the colour difference may distinguish species.

Bucca – the jowls on a dipteran head.

Caecum – a structure which ends in a blind end or in a sac, e.g. a region of the insect mid-gut.

Calypter – flap of tissue (often coloured white and with black hairs). There are usually two flaps at the base of each wing and these give the calyptrates their distinctive name (amongst other features). These flaps can also be called squamae.

Campodeiform – elongated and flattened, with clearly developed legs and antennae.

Cerci (singular **cercus**) – sensory appendages located at the tip of the abdomen.

Chromosome – a single DNA molecule which is a component of the genetic make-up of the organism and is morphologically apparent during cell division.

Cocaine – white crystalline alkaloid ($C_{17}H_{21}NO_4$); a narcotic.

Coleoptera – the order of insects comprising the beetles.

Coxae (singular **coxa**) – the proximal segments of the insect legs, i.e. at the base of the legs, where they join to the body.

CRFP – Council for the Registration of Forensic Practitioners.

Decomposition – organic matter breakdown into its constituents, in the case of the corpse by cell lysis and also microorganisms.

Dermal – to do with the skin, where the insects in the case of myiasis are consuming the skin tissue.

Diptera – The order comprising the flies.

Distal – distant from the body.

Dorsum – the upper top surface, and a term often used when referring to the thorax.

Dorsal – upper, top surface.

Dorsocentral – usually referring to bristles on the surface of the thorax. They are found to the sides of the acrostichal bristles.

Electrophoresis – where an electric current is applied to a gel loaded with molecules such as nucleic acids, to separate them.

Elytra (singular **elytron**) – the modified, hardened front wings of the coleopteran. The elytra cover the hind, membranous wings that are used for flight, where these are present.

Exarate – with free appendages. This may, for example, relate to the pupal stage of the life cycle.

Exoskeleton – the hardened, cuticular, external skeleton to which muscles are attached.

Exuvia (plural **exuviae**) – outer insect cuticle 'skin' shed at the end of each stadium.

Facultative – optional behaviour, where an organism is able to adapt its behaviour to take advantage of a particular situation or food source.

Filiform – thread-like; often used to describe the structure of the antennae.

Frass – solid insect, most often larval, waste products from digestion (another name for insect 'faeces').

Frons – the area between the antennae, eyes and running to the top of the head of the fly.

Furuncular – relating to boils.

Gena – the cheek of the insect (the region below and behind the insect eye).

Gendarmerie – a police force; often French or from a French-speaking country.

Gene – inherited genetic material, which is found in a fixed position, or locus, on a chromosome. It is part of the 'programming' for the production of a polypeptide.

Genome – the genetic composition of an organism.

Haltere – reduced hind wings used for balancing. They are sometimes described as resembling drumsticks, as they have a knob at the distal region.

Hemimetabolous – insects in which immature stages go through moults, gradually becoming increasingly like the adults. Their wings develop externally.

Heteroplasmy – two or more different mitochondrial genomes in the same species.

Holometabolous – insects which undergo complete metamorphosis.

Holarctic – the zoogeographical region which includes the northern regions of the earth. It has two sections, one called the nearctic region (North America) and the other the palearctic region (Eurasia). The terms are used to describe insect distributions.

Hypopleuron – a plate above the hind coxa and posterior to the stenopleuron (a plate, or pleuron, above the middle coxa). In *Calliphora* spp. there will be a crescent of bristles which are useful aids to identification.

Imago (plural **imagines**) – the adult stage of an insect.

Instar – the stage between two successive splittings of the outer layers of the larval cuticle (exoskeleton) from the inner layers (endoskeleton).

KAAD – a solution for killing larval specimens, containing ethanol, formaldehyde and kerosene in the proportion 10:2:1.

Kahle's solution – a solution for preserving dead larvae and for killing and preserving adult insects. The components are defined in the Appendix.

Larvae (singular **larva**) – collective term for the immature stages which emerge from the egg and prior to pupation.

Lateral – on or at the sides.

Mandibles – regions of the insect mouthparts; the jaw-like structures used for biting, e.g. in the Coleoptera.

Lysis – the breakdown or splitting of cells by enzymes. This takes place spontaneously after death.

Metabolism – the sum of the biochemical processes which take place in a living cell or organism.

Meron – the posterior lobe of the pleuron where the coxa joins to the side of the insect body.

Mesothorax – the second segment of the thorax.

Metamorphosis – the specific change in body form. The term can be applied, for example, to the transition between pupa and adult.

Metasternum – the under (ventral) side of the mesothorax.

Mitochondria (singular **mitochondrion**) – a rod-like organelle found in the cell cytoplasm in which energy is produced. Mitochondria contain circular DNA which can be duplicated. They originate from the maternal parent.

Monophyletic – descended from a single ancestor.

Myiasis – injury or secondary infection caused by larvae, usually fly larvae (Diptera), feeding on living human and animal tissue.

Myxophaga – a suborder of the Coleoptera which includes the Calyptomeridae. This suborder is of little forensic note.

Neotropical – a region which includes Central and South America, including southern Mexico, together with the West Indies, i.e. south of the Tropic of Cancer.

Notopleuron – the plate (pleurite) on the side of the fly body just at the end of the transverse suture.

Notum – a thoracic top surface of a segment (tergite).

Omnivores – organisms which can feed on a wide range of food materials; in this context they feed on the corpse and the insects present on it.

Opiates – a narcotic drug which contains opium, or an alkaloid of opium.

Order – a component of the classification system. There are 29 orders of insects, of which the Diptera and Coleoptera are two.

Orient – countries which are east of the Mediterranean.

PCR – Polymerase chain reaction. The cycles of denaturing the DNA, annealing the primer and extending with the enzyme DNA polymerase which result in increase in the amount of the target sequence of DNA.

Pedicel – the second antennal segment (working from the head outwards).

Pleuron (pleural **pleura**) – the side plates of the body, where the legs are joined to the sides of the body.

Polyphaga – a suborder of the Coleoptera characterized by moveable hind coxae, with respect to the metasternum, and an incomplete division of the first abdominal sternite.

Predator – an organism which consumes live organisms as its source of food, e.g. the Staphylinidae.

Primers – a short DNA (or RNA) sequence which is paired with one of the DNA strands from the test organism. An available 3′ hydroxyl group provides an anchor point for DNA replication to start.

Probative value – significance in terms of providing evidence or proof.

Prognathous – where the head points forward, i.e. is horizontal, and the mouthparts point to the front.

Prolegs – appendages on the abdominal region of the insect larval body.

Pronotum – first part of the upper plate of the thorax.

Prothorax – the first segment of the thorax of the insect body.

Predator – an insect which feeds off living organisms. In the context of forensic entomology, predators such as the Staphylinidae feed on those organisms attracted to the corpse.

Proximal – close to the body.

Ptilinum – a balloon-like sac pushed out from the head to force open the puparium and assist the emergence of the adult fly.

Ptilinal suture – the retracted ptilinum once the fly has emerged and the body hardened. The ptilinal suture is characteristic of flies of the Schizophora.

Pulvilli (singular **pulvillus**) – the two pads between the claws at the end of the tarsae.

Puparium – the final coat of the larval instar which becomes hardened and 'tanned'. Inside this casing the pupa develops into the adult stage in the Cyclorrapha.

RAPD – Random amplification of polymorphic DNA. Variable bands of DNA produced on a gel after PCR amplification.

RFLP – Restriction fragment length polymorphisms. Base changes at sites as a result of restriction digestion, resulting in different length DNA fragments.

Scape – first antennal segment.

Sclerotized – made up of a 'tanned protein' (sclerotin) which makes a hard, horny, outer layer of the cuticle. The structure could also be called 'chitinized'.

Scutum – the middle division of the top surface of a thoracic segment.

Sequester – to set aside, e.g. in fat cells; it is a means of storage or protection.

Species – a group of individual organisms which can interbreed and their offspring are fertile and resemble the parents.

Stadium (plural **stadia**) – the stage of morphological development between two moultings, e.g. L1 and L2.

Sterna (singular **sternite**) – ventral plates of the segments of the body (each individual plate is termed a sternite).

Sternum – the underside (ventral segmented region) of the insect thorax or abdomen.

Squamae (singular **squama**) – flaps of wing tissue (often coloured white and with black hairs). There are usually two such flaps at the base of each wing (also sometimes called calypters) and these (amongst other features) give the calyptrates their distinctive name.

Subdermal – below the skin.

Surstyli – sclerotized organs which are in pairs and together make up the lower parts of the male genitalia.

Suture – a 'seam' visible on the surface of the insect body which indicates where two plates join, e.g. the transverse suture running across the mesonotum in the Diptera.

Synanthropic – associated with humans and their activities.

Taxon (plural **taxa**) – these are over-arching terms for groups to which organisms are assigned, based upon principles of taxonomy, e.g. phylum, class, genus, species and so on.

Terga (singular **tergite**)– the top (upper or dorsal) plates of a segment, e.g. of the abdomen. An individual plate of any upper surface of the insect body is called a tergite.

Terrestrial – living or growing on the land, rather than in fresh or sea water.

Tessellated – chequered. This is a common description of the abdomen in flesh flies.

Thanatosis – appearing to be lifeless insect carcasses; 'playing dead'.

Tibia – the leg segment between the femur and the tarsus.

Urogomphi – small hardened structures projecting from the end of the larval abdomen.

Ventral – from below; underneath.

Vestigial – structures which have, over time, become degenerate or reduced.

Vertex – the top of the head.

Vibrissae – large bristles located at the sides of the mouth in some fly species.

Viviparous – females which produce live young rather than laying eggs. The Sarcophagidae are an example of such viviparous Diptera.

References

Adams Z. J. O. and Hall M. J. R. 2003. Methods used for the killing and preservation of blowfly larvae and their effect on post mortem length. *Forensic Science International* **138**(1–3): 50–61.

Anderson G. S. 1995. The use of insects in death investigations: analysis of cases in British Columbia over a five year period. *Canadian Society for Forensic Science Journal* **28**: 277–292.

Anderson G. S. 2000. Minimum and maximum development rates in some forensically important Calliphoridae (Diptera). *Journal of Forensic Sciences* **45**(4): 824–832.

Anderson G. S. and VanLaerhoven S. L. 1996. Initial studies on insect succession on carrion in southwestern British Columbia. *Journal of Forensic Sciences* **41**(4): 617–625.

Anderson G. S. 2005. Effects of arson on forensic entomology evidence. *Journal of the Canadian Society for Forensic Science* **38**(2): 49–67.

Archer M. S. 2004. The effect of time after body discovery on the accuracy of retrospective weather station ambient temperature corrections in forensic entomology. *Journal of Forensic Sciences* **49**(3): 1–7.

Archer M. S. and Elgar M. A. 1998. Cannibalism and delayed pupation in hide beetles, *Dermestes maculatus* DeGeer (Coleoptera: Dermestidae). *Australian Journal of Entomology* **37**(2): 158–161.

Archer M. S. and Elgar M. A. 1999. Female preference for multiple partners: sperm competition in the hide beetle, *Dermestes maculatus* (DeGeer). *Animal Behaviour* **58**(3): 669–675.

Archer M. S. and Elgar M. A. 2003. Yearly activity patterns in southern Victoria (Australia) of seasonally active carrion insects. *Forensic Science International* **132**(3): 173–176.

Arhuzhonov A. M. 1963. The use of entomological observations in forensic science *Sudebno Meditsinskaya Ékspertisa* **6**: 51 [in Russian].

Arnaldos M. I., Garcia M. D., Romera E., Presa J. J. and Luna A. 2005. Estimation of post mortem interval in real cases based on experimentally obtained entomological evidence *Forensic Science International* **149**(1): 57–65.

Austen E. E. 1910. Some dipterous insects which cause myiasis in man. *Transactions of the Society of Tropical Medicine and Hygiene* **3**(5): 215–242.

Avila F. A. and Goff M. L. 1998. Arthropod succession patterns on burnt carrion in two contrasting habitats in the Hawaiian islands. *Journal of Forensic Sciences* **43**(3): 581–586.

Avise J. C. 1991. Ten unorthodox perspectives on evolution prompted by comparative population genetic findings on mitochondrial DNA. *Annual Review of Genetics* **25**: 45–69.

Avise J. C. Arnold J., Ball R. M., Bermingham E. *et al.* 1987. Intraspecific phylogeography: the mitochondrial DNA bridge between population genetics and systematics. *Annual Review of Entomology* **18**: 489–522.

Barreto M., Burbano M. E. and Barreto P. 2002. Flies (Calliphoridae, Muscidae) and beetles (Silphidae) from human cadavers in Cali, Colombia (Short Communication). *Memorias do Instituto Oswaldo Cruz, Rio de Janeiro* **97**(1): 137–138.

Bartlett J. 1987. Evidence for a sex attractant in burying beetles. *Ecological Entomology* **12**: 471–472.

Baumgartner D. L. and Greenberg B. 1984. The genus *Chrysomya* (Diptera: Calliphoridae) in the New World. *Journal of Medical Entomology* **21**: 105–113.

Beard C. B., Hamm D. M. and Collins F. H. 1993. The mitochondrial genome of the mosquito *Anopheles gambiae*: DNA sequence, genome organisation and comparisons with the mitochondrial sequences of other insects. *Insect Molecular Biology* **2**: 103–124.

Benecke M. 1998. Random amplified polymorphic DNA (RAPD) typing of necrophageous insects (Diptera, Coleoptera) in criminal forensic studies: validation and use in practice. *Forensic Science International* **98**: 157–168.

Benecke M. 2001. A brief history of forensic entomology. *Forensic Science International* **120**(1–2): 2–14.

Benecke M. 2004. Arthropods and corpses. In Tsokos M. (ed.), *Forensic Pathology Reviews*, vol 2. Humana: Totowa, NJ; pp 207–240.

Benecke M. and Lessig R. 2001. Child neglect and forensic entomology. *Forensic Science International* **120**: 155–159.

Benecke M. and Barksdale L. 2003. Distinction of bloodstain patterns from fly artefacts. *Forensic Science International* **137**: 152–159.

Bergeret M. 1855. Infanticide. Momification naturelle du cadaver. *Annuals of Hygiene and Legal Medicine* **4**: 443–452.

Bhuiyan A. I. and Saifullah A. S. 1997. Biological note on *Necrobia rufipes* (Deg.) (Coleoptera: Cleridae). *Bangladesh Journal of Zoology* **25**: 121–124.

Block W., Erzinçlioğlu Y. Z. and Worland M. R. 1990. Cold resistance in all life stages of two blowfly species (Diptera, Calliphoridae). *Medical and Veterinary Entomology* **4**: 213–219.

Bornemissza G. F. 1957. An analysis of arthropod succession in carrion and the effect of its decomposition on the soil fauna. *Australian Journal of Zoology* **5**: 1–12.

Bourel B., Hédouin V., Martin-Bouyer L., Bécart A. *et al.* 1999. Effects of morphine in decomposing bodies of *Lucilia sericata* (Diptera: Calliphoridae). *Journal of Forensic Sciences* **44**(2): 354–358.

Bovingdon H. H. S. 1933. Report on the infestation of cured tobacco in London by the Cacao moth *Ephestia elutella* Hb. Empire Marketing Board 67. HMSO: London.

Brandt A. 2004. Insect activity on pig carcasses over winter in London. Proceedings of European Association for Forensic Entomology Conference, 29–30 March 2004, London; p 42.

Brisard C., 1939 *Pediculus vestimenti. Annales de Médicine Légale de Criminologie et le Police Scientifique* **9–10**: 614–615.

Broadhead E. C. 1980. Larvae of trichocerid flies found on human corpse. *Entomologist's Monthly Magazine* **116**: 23–24.

Busvine J. R. 1980. *Insects and Hygiene*, 3rd edn. Chapman and Hall: London.

Byrd J. H. and Butler J. F. 1998. Effects of temperature on *Sarcophaga haemorrhoidalis* (Diptera: Sarcophagidae) development. *Journal of Medical Entomology* **35**(5): 694–698.

Byrd J. and Allen J. C. 2001. The development of the black blowfly, *Phormia regina* (Meigen). *Forensic Science International* **120**(1–2): 79–88.

Byrd J. H. and Castner J. L. (eds). 2001. *Forensic Entomology: The Utility of Arthropods in Legal Investigations.* CRC Press: Boca Raton, FL.

Byrne A. L., Camann M. A., Cyr T. L., Catts E. P. and Espelie K. E. 1995. Forensic implications of biochemical differences among geographic populations of the black blowfly, *Phormia regina* (Meigen). *Journal of Forensic Sciences* **40**(3): 372–377.

Campan M., Le Pape G. and Benziane T. 1994. Description du comportement sexuel de *Calliphora vomitoria* (Diptera: Calliphoridae) par une technique d'analyse de texts. *Behavioural Processes* **31**(2–3): 269–284.

Campobasso C. P., Di Vella G. and Introna F. 2001. Factors affecting decomposition and Dipteran colonization. *Forensic Science International* **120**: 18–27.

Campobasso C. P., Linville J. G., Wells J. D. and Introna F. M. D. 2005. Forensic genetic analysis of insect gut contents. *American Journal of Forensic Medicine and Pathology* **26**(2): 161–165.

Carvalho L. M. I., Thyssen P. J., Linhares A. X. and Palhares F. A. B. 2000. A checklist of arthropods associated with pig carrion and human corpses in south-eastern Brazil. *Memorias do Instituto Oswaldo Cruz* **195**(1): 135–138.

Catts E. P. 1992. Problems in estimating the post mortem interval in death investigations. *Journal of Agricultural Entomology* **9**(4): 245–255.

Catts E. P. and Haskell N. H. (eds). 1990. *Entomology and Death: A Procedural Guide.* Joyce's Print Shop: Clemson, SC.

Centeno N., Maldonado M. and Oliva A. 2002. Seasonal patterns of arthropods occurring on sheltered and unsheltered pig carcasses in Buenos Aires Province (Argentina). *Forensic Science International* **126**(1): 63–70.

Chandler P. J. 1998. Checklists of Insects of the British Isles (New Series), Part 1: Diptera. *Handbooks for the Identification of British Insects*, vol 12. Royal Entomological Society: London.

Chapman R. F. and Sankey J. H. P. 1955. The larger invertebrate fauna of three rabbit carcasses. *Journal of Animal Ecology* **24**: 395–402.

Chen Chun-Hsien and Shih Cheng-Jen. 2003. Rapid identification of three species of blowflies (Diptera: Calliphoridae) by PCR-RFLP and DNA sequencing analysis. *Formosan Entomology* **23**: 59–70 [in Chinese with English abstract and figure titles].

Cheng Ko. 1890. Cases in the history of Chinese trials [English translation of Zhe yu gui jian bu] Lu Shih China. In *Entomology and the Law*, Greenberg B. and Kunich J.C. (eds). Cambridge University Press: Cambridge, 2002.

Chinnery M. 1973. *A Field Guide to the Insects of Britain and Northern Europe.* William Collins: London.

Clarke T. E., Levin D. B., Kavanaugh D. H. and Reimchen T. E. 2001. Rapid evolution in the *Nebria Gregaria* group (Coleoptera: Carabidae) and the palaeography of the Queen Charlotte Islands. *Evolution* **55**(7): 1408–1418.

Clary D. O. and Wolstenholme D. R. 1985. The mitochondrial DNA molecule of *Drosophila yakuba*: nucleotide sequence, gene organisation and genetic code. *Journal of Molecular Evolution* **22**: 252–271.

Coffey M.D. 1966. Studies on the association of flies (Diptera) with dung in south-eastern Washington. *Annals of the Entomological Society of America* **59**: 207–218.

Colyer C. N. 1954. The 'coffin' fly, *Conicera tibialis* Schmidtz (Dipt., Phoridae). *Journal of the Society for British Entomology* **4**(9): 203–206.

Colyer C. N. and Hammond C. O. 1951. *Flies of the British Isles.* Frederick Warne: London.

Conquest E. 1999. The pheromone-mediated behaviour of *Dermestes maculatus.* Poster presentation at the International Society of Chemical Ecology Meeting, 13–17 March 1999, Marseille.

Coombs C. W. 1978. The effect of temperature and relative humidity upon development and fecundity of *Dermestes lardarius* L. (Coleoptera, Dermestidae). *Journal of Stored Product Research* **14**: 111–119.

Cooter J. 2006. Cleroidea. In Cooter J. and Barclay M. V. L., *A Coleopterist's Handbook*, 4th edn. Amateur Entomologists' Society: Orpington, Kent, UK.

Cooter J. and Barclay M. V. L. 2006. *A Coleopterist's Handbook*, 4th edn. Amateur Entomologists' Society: Orpington, Kent, UK.

Cragg J. B. 1955. The natural history of sheep blowflies in Britain. *Annals of Applied Biology* **42**: 197–207.

Cragg J. B. 1956. The olfactory behaviour of *Lucilia* species (Diptera) under natural conditions. *Annals of Applied Biology* **44**: 467–477.

Crosby T. K., Watt J. C., Kistemaker A. C. and Nelson P. E. 1986. Entomological identification of the origin of imported cannabis. *Journal of the Forensic Science Society* **26**: 35–44.

Crowson R. A. 1981. *The Biology of the Coleoptera.* Academic Press: London.

Davies L. 1990. Species composition and larval habitats of blowfly (Calliphoridae) populations in upland areas of England and Wales. *Medical and Veterinary Entomology* **4**: 61–68.

Davies L. 1998. Delayed egg production and a possible group effect in the blowfly *Calliphora vicina. Medical and Veterinary Entomology* **12**: 339.

Davies L. and Ratcliffe G. G. 1994. Development rates of some pre–adult stages in blowflies with reference to low temperatures. *Medical and Veterinary Entomology* **8**: 245–254.

Dean M. D. and Ballard J. W. O. 2001. Factors affecting mitochondrial DNA quality from museum preserved *Drosophila simulans. Entomologia Experimentalis et Applicata* **98**: 279–283.

Dear J. P. 1978. Carrion. In Stubbs A. and Chandler P. (eds), *A Dipterist's Handbook. The Amateur Entomologist* **15.** The Amateur Entomologists' Society: Orpington.

DeFoliart G. R. 1988. Query: are processed insect food products still commercially available in the United States? *The Food Insects Newsletter* **1**(2, November): 1.

de Hough G. N. 1897. The fauna of dead bodies, with especial reference to Diptera. *British Medical Journal* 1853–1854.

Deonier C. C. 1942. Seasonal abundance and distribution of certain blowflies in southern Arizona and their economic importance. *Journal of Economic Entomology* **35**: 65–70.

Dewaele P. and LeClerq M. 2002. Les Phorides (Diptères) sur cadavers humains en Europe occidentale. Proceedings of the First European Forensic Entomology Seminar 2002; pp 79–86.

Digby P. S. B. 1958. Flight activity of the blow fly, *Calliphora erythrocephala*, in relation to wind speed, with special reference to adaptation. *Journal of Experimental Biology* **35**: 776–795.

Dillon L. and Anderson G. S. 1996. Forensic entomology: the use of insects in death investigations to determine elapsed time since death in interior and northern British Columbia regions. Technical report TR–03–96. Canadian Police Research Centre: Ottawa, Ontario.

Dillon N., Austin A. D. and Bartowsky E. 1996. Comparison of preservation techniques for DNA extraction from hymenopterous insects. *Insect Molecular Biology* **5**(1): 21–24.

Disney R. H. L. 2005. Duration of development of two species of carrion breeding scuttle flies and the forensic implications. *Medical and Veterinary Entomology* **19**(2): 229–235.

Disney R. H. L. and Manlove J. D. 2005. First occurrences of the phorid, *Megaselia abdita*, in forensic cases in Britain. *Medical and Veterinary Entomology* **19**(4): 489–491.

Di Zinno J. A., Lord W. D., Collins-Morton M. B., Wilson M. R. and Goff M. L. 2002. Mitochondrial DNA sequencing of beetle larvae (Nitidulidae: *Omosita*) recovered from human bone. *Journal of Forensic Sciences* **47**(6): 1337–1339.

Dobler S. and Muller J. K. 2000. Resolving phylogeny at the family level by mitochondrial cytochrome oxidase sequences: phylogeny of carrion beetles (Coleoptera, Silphidae). *Molecular Phylogenetics and Evolution* **15**(3): 390–402.

Donovan S. E., Hall M. J. R., Turner B. D. and Moncrieff C. B. 2006. Larval growth rates of the blowfly, *Calliphora vicina*, over a range of temperatures. *Medical and Veterinary Entomology* **10.111/j**; 1365–2915. 2006.00600.x.

Durden L. A. 2002. Lice. In Mullen G. and Durden L (eds), *Medical and Veterinary Entomology*. Academic Press: Amsterdam; pp 45–65.

Easton A. M. and Smith K. G. V. 1970. The entomology of the cadaver. *Medicine, Science and the Law* **10**: 208–215.

Edwards F. W. 1928. Diptera Fam. Protorhyphidae, Anisopodidae, Pachyneuridae, Trichoceridae (with descriptions of early stages by D. Keilin). *Gen. Insect* **190**: 1–41. In Keilin D. and Tate P. 1940. The early stages of the families Trichoceridae and Anisopodidae (= Rhyphidae) (Diptera: Nematocera). *Transactions of the Royal Entomological Society of London* **90**(3): 39–62.

Eggert A.-K., Reinking M. and Muller J. K. 1998. Parental care improves offspring survival and growth in burying beetles. *Animal Behaviour* **55**: 97–107.

Erzinçlioğlu Y. Z. 1980. On the role of Trichoceridae larvae (Diptera: Trichoceridae) in decomposing carrion in winter. *Naturalist* **105**: 133–134.

Erzinçlioğlu Y. Z. 1987. Recognition of the early instar larvae of the genera *Calliphora* and *Lucilia* (Diptera: Calliphoridae). *Entomologist's Monthly Magazine* **123**: 97–98.

Erzinçlioğlu Y. Z. 1996. *Blowflies*. Naturalists' Handbooks No. 23. Richmond Publishing: Slough.

Erzinçlioğlu Y. Z. 2000. *Maggots, Murder and Men*. Harley Press: Colchester.

Faria Del Bianco L., Orsi L., Trinca L. A. and Godoy W. A. C. 1999. Larval predation by *Chrysoma albiceps* on *Cochliomyia macellaria*, *Chrysomya megacephala* and *Chrysomya putoria*. *Entomologia Experimentali et Applicata*. **90**(2): 149–157.

Faucherre J., Cherix D. and Wyss C. 1999. Behaviour of *Calliphora vicina* (Diptera, Calliphoridae) under extreme conditions. *Journal of Animal Behaviour* **12**: 687–690.

Fisher P., Wall R. and Ashworth J. R. 1998. Attraction of the sheep blowfly, *Lucilia sericata* (Diptera: Calliphoridae) to carrion bait in the field. *Bulletin of Entomological Research* **88**: 611–616.

Fraenkel G. 1935. Observations and experiments on the blowfly (*Calliphora erythro-cephala*) during the first day after emergence. *Proceedings of the Zoological Society of London* 1935: 893–904.

Fukatsu T. 1999. Acetone preservation: a practical technique for molecular analysis *Molecular Ecology* **8**: 1935–1945.

Gaudry E. 2002. Eight squadrons for one target: the fauna of cadaver described by P. Mégnin. Proceedings of the First European Forensic Entomology Seminar, 77 eafe.org/OISIN_2002; p 23.

Gaudry E., Myskowiak J.-B., Chauvet B., Pasquerault T *et al.* 2001. Activity of the forensic entomology department of the French Gendarmerie. *Forensic Science International* **120**(1–2): 68–71.

Gaudry E., Dourel L., Zehner R. and Amendt J. 2004. Quality assurance in forensic entomology: why, how and who? Proceedings of the European Association for Forensic Entomology Conference, 29–30 March 2004, London; p 21.

Gennis R. B. 1992. Site-directed mutagenesis studies of subunit I of the aa3-type cytochrome *c* oxidase of *Rhodobacter sphaeroides*: a brief review of progress to date. *Biochemica et Biophysica Acta* **11010**: 184–187.

Giertsen J. C. 1977. Drowning in forensic medicine. In Tedeschi C. G., Eckert W. G. and Tedeschi L. G. (eds), *Forensic Medicine: A Study in Trauma and Environmental Hazards*, vol III. Saunders: Philadelphia, PA; pp 1317–1333.

Gill G. J. 2005. Decomposition and arthropod succession on above ground pig carrion in rural Manitoba. Technical report TR–06–2005. Canadian Police Research Centre: Ottawa, Ontario.

Gilmour D., Waterhouse D. F. and McIntyre K. L. 1946. An account of experiments undertaken to determine the natural population of sheep blowfly, *Lucilia cuprina* Weid. *Bulletin of the Council of Scientific and Industrial Research Australia* **195**: 1–39.

Glassman D. M. and Crow R. M. 1996. Standardization model for describing the extent of burn injury to human remains. *Journal of Forensic Sciences* **41**(1): 152–154.

Goff M. L. 1993. Estimation of post mortem interval using arthropod development and successional patterns. *Forensic Science Review* **5**(2): 81–94.

Goff M. L. 1998. Arthropod succession patterns on burnt carrion in two contrasting habitats in the Hawaiian islands. *Journal of Forensic Sciences* **43**(3): 581–586.

Goff M. L. 2000. *A Fly for the Prosecution.* Harvard University Press: Cambridge, MA.

Goff M. L., Brown W. A., Hewadikaram K. A. and Omori A. I. 1991. Effect of heroin in decomposing tissues on the development rate of *Boettcherisca peregrina* (Diptera, Sarcophagidae) and implications of this effect on estimation of post mortem intervals using arthropod development rates. *Journal of Forensic Sciences* **36**(2): 537–542.

Goff M. L. and Flynn M. F. 1991. Determination of post mortem interval by arthropod succession: a case study from the Hawaiian Islands. *Journal of Forensic Sciences* **36**: 607–614.

Grassberger M. and Reiter C. 2001. Effect of temperature on *Lucilia sericata* (Diptera: Calliphoridae) development with special reference to isomegalen- and isomorphen-diagram. *Forensic Science International* **120**(1–2): 32–36.

Grassberger M. and Reiter C. 2002. Effect of temperature on development of the foren-sically important holarctic blowfly *Protophormia terraenovae* (Robineau-Desvoidy) (Diptera: Calliphoridae). *Forensic Science International* **128**: 177–182.

Grassberger M., Freidrich E. and Reiter C. 2003. The blowfly *Chrysomya albiceps* (Weidmann) (Diptera: Calliphoridae) as a new forensic indicator in Central Europe. *International Journal of Legal Medicine* **117**: 75–81.

Grassberger M. and Frank C. 2004. Initial study of arthropod succession on pig carrion in a central European urban habitat. *Journal of Medical Entomology* **41**(3): 511–523.

Green A. A. 1951. The control of blowflies infesting slaughterhouses 1. Field observation on the habits of blowflies. *Annals of Applied Biology* **38**: 475–494.

Greenberg B. 1990. Nocturnal oviposition behaviour of blowflies (Diptera: Calliphoridae). *Journal of Medical Entomology* **27**(5): 807–810.

Greenberg B. 1991. Flies as forensic indicators. *Journal of Medical Entomology* **28**: 565–577.

Greenberg B. and Singh D. 1995. Species identification of calliphorid (Diptera) eggs. *Journal of Medical Entomology* **32**(1): 21–26.

Greenberg B. and Kunich J. C. 2002. *Entomology and the Law: Flies as Forensic Indicators.* Cambridge University Press: Cambridge.

Grimshaw P. H. 1917. A guide to the literature of British Diptera. *Proceedings of the Royal Physical Society Edinburgh* **20**: 78–117.

Grimshaw P. H. 1934. Introduction to the study of Diptera, with a key to the identification of families. *Proceedings of the Royal Physical Society Edinburgh* **22**: 187–215.

Gunatilake K. and Goff M. L. 1989. Detection of organophosphate poisoning in a putrefying body by analysing arthropod larvae. *Journal of Forensic Sciences* **34**: 714–716.

Hakbijl T. 2000. Arthropod remains as indicators for taphonomic processes: an assemblage from 19th century burials, Broerenkerk, Zwolle, The Netherlands. In Huntley J. P. and Stallibrass S. (eds), *Taphonomy and Interpretation.* Symposia for the Association for Environmental Archaeology, No 14. Oxbow Books: Oxford; pp 95–96.

Hall D. G. 1948. *The Blowflies of North America.* Say: Baltimore, MD.

Hall M. J. R. and Smith K. G. V. 1993. Diptera causing myiasis in man. In Lane R. P. and Crosskey R. W. (eds), *Medical Insects and Arachnids.* Chapman and Hall: London; pp 429–469.

Hanks P. 1984. *The Collins Dictionary of the English Language.* Collins: London.

Hanski I. 1987. Carrion fly community dynamics: patchiness, seasonality and coexistence. *Ecological Entomology* **12**: 257–266.

Harrison D. A., Cooper R. L. 2003. Characterization of development, behaviour and neuromuscular physiology in the phorid fly, *Megaselia scalaris. Comparative Biochemistry and Physiology A* **136**: 427–439.

Harvey M. L., Dadour I. R. and Gaudieri S. 2003. Mitochondrial DNA cytochrome oxidase I gene: potential for distinction between immature stages of forensically important fly species (Diptera) in western Australia. *Forensic Science International* **131**: 134–139.

Haskell N. H. 2000. Testing reliability of animal models in forensic entomology with 50–200 lb pig vs. humans in Tennessee. Proceedings of the XXI International Congress on Entomology, Brazil; abstr 2943.

Hédouin V., Bourel B., Bécart A., Gilles D. D. S. *et al.* 2001. Determination of drugs levels in larvae of *Protophormia terraenovae* and *Calliphora vicina* (Diptera: Calliphoridae) reared on rabbit carcasses containing morphine. *Journal of Forensic Sciences* **46**(1): 12–14.

Hedström L. and Nuorteva P. 1971. Zonal distribution of flies on the hill Aiiglas in sub-arctic northern Finland. *Annals Entomologici Fennici* **37**(2): 121–125.

Hermes V. B. 1928. The effect of different quantities of food during the larval period on the sex ratio of *Lucilia sericata* Meigen and *Theobaldia incidens* (Thom.). *Journal of Economic Entomology* **21**: 720–729.

Higley L. G. and Haskell N. H. 2001. Insect development and forensic entomology. In Byrd J. H. and Castner J. L. (eds), *Forensic Entomology: The Utility of Arthropods in Legal Investigations.* CRC Press: Boca Raton, FL.

Hinton H. E. 1945. *A Monograph of the Beetles Associated with Stored Products*, vol 1. Trustees of the Natural History Museum: London.

Hinton H. E. 1981. *The Biology of Insect Eggs.* Pergamon Press: Oxford.

Hoback W. W., Bishop A. A., Kroemer J., Scalzitti J. M. and Shaffer J. J. 2004. Differences among the antimicrobial properties of carrion secretions reflect phylogeny and ecology. *Journal of Chemical Ecology* **30**(4): 719–729.

Hobischak N. R. and Anderson G. S. 2002. Time of submergence using aquatic invertebrate succession and decompositional changes. *Journal of Forensic Sciences* **47**(1): 142–151.

Hough G. de N. 1897. The fauna of dead bodies with especial reference to Diptera. *British Medical Journal*: 1853–1854.

Huijbregts H. 2004. Distribution of the blowflies, *Phormia regina* and *Protophormia terraenovae* in The Netherlands and Western Europe. Proceedings of the European Association for Forensic Entomology Conference, 29–30 March 2004, London; p 5.

Iannacone J. 2003. Arthropofauna of forensic importance in pig carcass in Callao, Peru. *Revista Brasileira de Zoologia* **20**: 85–90.

Introna F., Altamura B. M., Dell'Erba A. and Dattoli V. 1989. Time since death definition by experimental reproduction of *Lucilia sericata* cycles in a growth cabinet. *Journal of Forensic Sciences* **34**(2): 478–480.

Introna F., Lo Dico C., Caplan Y. H. and Smialek J. E. 1990. Opiate analysis in cadaveric blowfly larvae as an indicator of narcotic intoxication. *Journal of Forensic Sciences* **35**: 118–122.

Introna F., Suman T. W., Smialek J. E. 1991 Sarcosaprophagous fly activity in Maryland. *Journal of Forensic Sciences* **36**(1): 238–243.

James M.T. 1947. *The Flies that Cause Myiasis in Man.* Miscellaneous Publication 631. US Department of Agriculture: Washington, DC; pp 1–175.

Johnson C. and Esser J. 2000. *A Review of Insect Infestation of Traditionally Cured Fish in the Tropics.* Department of International Development: London; 92 pp.

Johnson W. and Villeneuve G. 1897. On the medico-legal application of entomology. *Montreal Medical Journal* **26**: 81–90.

Jones T. M. and Elgar M. A. 2004. The role of male age, sperm age and mating history on fertilisation success in the hide beetle *Dermestes maculatus. Proceedings of the Royal Society of London Series B* **271**: 1311–1318.

Junqueira A. C. M., Lessinger A. C. and Azendo-Espin A. M. L. 2002. Methods for the recovery of mitochondrial DNA sequences from museum specimens of myiasis-causing flies. *Medical and Veterinary Entomology* **16**: 39–45.

Kamal A. S. 1958. Comparative study of 13 species of sarcosaprophagous Calliphoridae and Sarcophagidae (Diptera). 1. Bionomics. *Annals of the Entomological Society of America* **51**: 261–270.

Kaneshrajah G. and Turner B. 2004. *Calliphora vicina* larvae grow at different rates on different body tissues. *International Journal of Legal Medicine* **118**(4): 242–244.

Keilin D. and Tate P. 1940. The early stages of the families Trichoceridae and Anisopodidae (=Rhyphidae) (Diptera: Nematocera). *Transactions of the Royal Entomological Society of London* **90**(3): 39–62.

Keiper J. B. and Casamatta D. A. 2001. Benthic organisms as forensic indicators. *Journal of the North American Benthological Society* **20**(2): 311–324.

Kintz P., Godelar B., Tracqui A., Mangin P. *et al.* 1990. Fly larvae: a new toxicological method of investigation in forensic science. *Journal of Forensic Sciences* **35**: 204–207.

Kloet G. S. and Hincks W. D. 1976. A check list of British insects. Part 5: Diptera and Siphonoptera, 2nd edn (completely revised). *Handbooks for the Identification of British Insects* **11**(5). Royal Entomological Society: London; 1–139.

Komar D. and Beattie O. 1998. Post mortem insect activity may mimic perimortem sexual assault clothing patterns. *Journal of Forensic Sciences* **43**(4): 792–796.

Korvarik P. W. 1995. Development of *Epierus divisus* Marseul (Coleoptera: Histeridae). *The Coleopterists' Bulletin* **49**(3): 253–260.

Kramer Wilson K. A. 1999. Parenting behaviour and long-term implications for reproduction success of burying beetles (Coleoptera: Silphidae: *Nicrophorus*): www.colostate.edu/depts/Entomology/courses/en507/papers-1999/kramer.htm

Kulshrestha P. and Satpathy D. K. 2001. Use of beetles in forensic entomology. *Forensic Science International* **120**: 15–17.

Lane R. P. 1975. An investigation into blowfly (Diptera: Calliphoridae) succession on corpses. *Journal of Natural History* **9**: 581–588.

Lawrence J. F. and Newton A. F. Jr. 1982. Evolution and classification of beetles. *Annual Review of Ecology and Systematics* **13**: 261–290.

Lefebvre F. and Pasquerault T. 2004. Temperature-dependent development of *Ophyra aenescens* (Weidemann, 1830) and *Ophyra capensis* (Weidemann, 1818). *Forensic Science International* **139**: 75–79.

Lessinger A. C., Junqueira Martins A. C., Lemos T. A., Kemper E. L. *et al.* 2000. The mitochondrial genome of the primary screwworm fly *Cochliomyia hominivorax* (Diptera: Calliphoridae). *Insect Molecular Biology* **9**: 521–529.

Levinson Z. H. 1962. The function of dietary sterols in phytophagous insects. *Journal of Insect Physiology* **8**(2): 191–198.

Levinson A. R., Levinson H. Z. and Franke D. 1981. Intraspecific attractants of the hide beetle, *Dermestes maculatus* Deg. *Mitteilungen der Deutschen Gesellschaft fur Allemeine* **2**: 235–237.

Levinson H. Z., Levinson A. R., Jen T.-L., Williams J. L. D. *et al.* 1978. Production site, partial composition and olfactory perception of a pheromone in the male hide beetle. *Naturwissernschaften* **10**: 543–545.

Linville J. G. and Wells J. D. 2002. Surfaces sterilisation of a maggot using bleach does not interfere with mitochondrial DNA analysis of crop contents. *Journal of Forensic Sciences* **47**(5): 1055–1059.

Lonsdale H. L., Dixon R. A. and Gennard D. E. 2004. Comparison of the efficiency of mitochondrial DNA extraction and assessment in aged and modern dipteran samples. Forensic Analysis 2004: Royal Society of Chemistry Conference, 20–22 June 2004, University of Lincoln.

Lord W., Catts E. P., Scarboro D. A. and Hadfield D. B. 1986. The green blowfly *Lucilia illustris* (Meigen) as an indicator of human post mortem interval: a case of homicide from Fort Lewis, Washington. *Bulletin of the Society of Vector Ecology* **11**: 271–275.

Lord W. D., Adkins T. R. and Catts E. P. 1992. The life cycle for *Synthesiomyia nudiseta* (Van der Wulp) (Diptera: Muscidae) and *Calliphora vicina* (Robineau-Desvoidy) (Diptera: Calliphoridae) to estimate the time of death of a body buried under a house. *Journal of Agricultural Entomology* **9**(4): 227–235.

Loxdale H. D. and Lushai G. 1998. Molecular markers in entomology. *Bulletin of Entomological Research* **88**: 577–600.

Luff M. 2006. Rearing larvae. In Cooter J. and Barclay M. V. L., *A Coleopterist's Handbook*, 4th edn. Amateur Entomologists' Society: Orpington, Kent, UK.

Lundt H. 1964 Ökologische untersuchungen über tierische beseidlung von Aas im Boden. *Pedobiologia* **4**: 158–180.

McManus J. J. 1974. Oxygen consumption in the beetle *Dermestes maculatus*. *Comparative Biochemistry and Physiology A* **49**(1): 169–173.

MacLeod J. and Donnelly J. 1956. Methods for the study of blowfly populations: the use of laboratory-bred material. *Annals of Applied Biology* **44**(4): 643–648.

MacLeod J. and Donnelly J. 1960. Natural features and blowfly movement. *Journal of Animal Ecology* **29**(1): 85–93.

Malgorn Y. and Coquoz R. 1999. DNA typing for identification of some species of Calliphoridae: an interest in forensic entomology. *Forensic Science International* **102**: 111–119.

Mann R. W., Bass W. M. and Meadows L. 1990. Time since death and decomposition of the human body: variables and observations in case and experimental field studies. *Journal of Forensic Sciences* **35**: 103–111.

Marchenko M. L. 2001. Medico-legal relevance of cadaver entomofauna for the determination of the time of death. *Forensic Science International* **120**(1–2): 89–109.

Meek L. 1990. Unpublished work, cited in Catts E. P. and Goff M. L. 1992. Forensic entomology in criminal investigations. *Annual Review of Entomology* **37**: 253–272.

Mégnin P. 1887. La faune des tombeaux. *Compte Rendu Hebdomadaire des Séances de l'Académie des Sciences* **105**: 948–951.

Mégnin P. 1894. *La Faune des Cadavres*. Encyclopédie Scientifiques des Aide Memoire. Masson, Gauthier-Villars et Fils: Paris.

Menzel P. and D'Aluisio F. 1998 *Men Eating Bugs: The Art and Science of Eating Insects*. Ten Speed Press: Berkeley, CA.

Merritt R. W. and Wallace J. R. 2001. The role of aquatic insects in forensic investigations. In Byrd J. H. and Castner J. L. (eds), *Forensic Entomology: The Utility of Arthropods in Legal Investigations*. CRC Press: Boca Raton, FL.

Miller M. L., Lord W. D., Goff M. L., Donnelly B. *et al.* 1994. Isolation of amitriptyline and nortriptyline from fly puparia (Phoridae) and beetle exuvae (Dermestidae) associated with mummified human remains. *Journal of Forensic Sciences* **39**(5): 1305–1313.

Morlais I. and Severson D. W. 2002. Complete mitochondrial DNA sequence and amino acid analysis of the cytochrome *c* oxidase subunit I (COI) from *Aedes aegypti*. *DNA Sequence* **13**(2): 123–127.

Mostovski M. and Mansell M. W. 2004. Life after death: true flies recover the truth. National Cemeteries and Crematoria Conference, 15–16 July 2004, Durban.

Munro J. W. 1966. *Pests of Stored Products*. The Rentokil Library. Benham and Co: Colchester.

Mullen G. and Durden L. 2002. *Medical and Veterinary Entomology*. Academic Press: Amsterdam.

Muller J. K. and Eggert A.-K. 1990. Time-dependent shifts between infanticidal and parental behaviour in female burying beetles: a mechanism of indirect mother–offspring recognition. *Behaviour, Ecology and Sociobiology* **27**: 11–16.

Musvaska E., Williams K. A., Muller W. J. and Villet M. H. 2001. Preliminary observations on the effects of hydrocortisone and sodium methohexital on development of *Sarcophaga* (*Curranea*) *tibialis* Macquart (Diptera: Sarcophagidae), and implications for estimating post mortem interval. *Forensic Science International* **120**: 37–41.

Myskowiak J.-B. and Doums C. 2002. Effects of refrigeration on the biometry and development of *Protophormia terraenovae* (Robineau-Desvoidy) (Diptera: Calliphoridae) and its consequences in estimating post-mortem interval in forensic investigations. *Forensic Science International* **125**: 254–261.

Nolte K. B., Pinder R. D. and Lord W. D. 1992. Insect larvae used to detect cocaine poisoning in a decomposed body. *Journal of Forensic Sciences* **37**: 1179–1185.

Nuorteva P. 1963. Synanthropy of blowflies (Dipt., Calliphoridae) in Finland. *Annals Entomologici Fennici* **29**: 1–49.

Nuorteva P. 1977. Sarcosaprophagous insects as forensic indicators. In Tedeschi C. G., Eckert W. G. and Tedeschi L. G. (eds), *Forensic Medicine: A Study in Trauma and Environmental Hazards*, vol II. Saunders: Philadelphia, PA; pp 1072–1095.

Nuorteva P. 1987. Empty puparia of *Phormia terraenovae* R–D (Diptera, Calliphoridae) as forensic indicators. *Annales Entomologici Fennici* **53**: 53–56.

Obsuji F. N. C. 1975. Some aspects of the biology of *Dermestes maculatus* De Geer (Coleoptera, Dermestidae) in dried fish. *Tropical Stored Product Information* **29**: 21–32.

Odeyemi O. O. 1997. Interspecific competition between the beetles *Dermestes maculatus* De Geer and *Necrobia rufipes* De Geer on dried fish. *Insect Science and Its Application* **17**: 213–220.

Oldroyd H. 1964. *The Natural History of Flies*. The World Naturalist Series. Weidenfield and Nicholson: London.

Oliva A. 2001. Insects of forensic significance in Argentina. *Forensic Science International* **120**(1–2): 145–154.

Oliveira-Costa J. and de Mello-Patiu C. A. 2004. Application of forensic entomology to estimate of the post-mortem interval (PMI) in homicide investigations by the Rio de Janeiro Police Department in Brazil. *Aggrawal's Internet Journal of Forensic Medicine and Toxicology* **5**(1): 40–44.

Patton W. S. 1922. Notes on myiasis-producing diptera of man and animals. *Bulletin of Entomological Research* **XII**(3): 239–261.

Payne J. A. 1965. A summer carrion study of the baby pig, *Sus scofa* Linnaeus. *Ecology* **46**(5): 592–602.

Payne J. A., King E. W. and Beinhart G. 1968. Arthropod succession and decomposition of buried pigs. *Nature* **219**(September 14): 1180–1181.

Peacock E. 1993. Adults and larvae of hide, larder and carpet beetles and their relatives (Coleoptera: Dermestidae) and of Derodontid beetles (Coleoptera: Derodontidae). *Handbooks for the Identification of British Insects* **5**(3). Royal Entomological Society: London.

Pont A. C. and Meier R. 2002. The Sepsidae (Diptera) of Europe. *Fauna Entomologica Scandinavica* **37**.

Prinkkilä M.-L. and Hanski I. 1995. Complex competitive interactions in four species of *Lucilia* blowflies. *Ecological Entomology* **20**: 261–272.

Pukowski E. 1933. Okologische untersuchungen an *Necrophorus* F. *Zeitschrift für Morphologie und Ökologie der Tiere* **27**: 518–586.

Rabinovich J. E. 1970. Vital statistics of *Synthesiomyia nudiseta* (Diptera: Muscidae). *Annals of the Entomological Society of America* **59**: 749–752.

Rakowski G. and Cymborowski B. 1982. Aggregation pheromone in *Dermestes maculatus*: effects on larval growth and developmental rhythms. *International Journal of Invertebrate Reproduction* **4**: 249–254.

Rankin M. R. and Bates P. G. 2004. Ageing blowfly strike lesions by larval morphology. Proceedings of the European Association of Forensic Entomology Conference, London: p 34.

Raspi A. and Antonelli R. 1995. Influence of temperature on the development of *Dermestes maculatus* Deg. (Coleoptera, Dermestidae). *Frustula Entomologica* **18**: 169–176.

Reiter C. 1984. Zum Wachstumsverhalten der Maden der blauen Schmeissfliege *Calliphora vicina*. *Zeitschrift fur Rechtsmedizen* **91**: 295–308.

Richards E. N. and Goff M. L. 1997. Arthropod succession on exposed carrion in three contrasting tropical habitats on Hawaii Island, Hawaii. *Journal of Medical Entomology* **34**(3): 328–339.

Richards O. W. and Davies R. G. 1988. *Imms' General Textbook of Entomology*, 10th edn, 2 vols. Chapman and Hall: London.

Rodriguez W. C. and Bass W. M. 1983. Insect activity and its relationship to decay rates in human cadavers in Eastern Tennessee. *Journal of Forensic Sciences* **28**(2): 423–432.

Rognes K. 1991. Blowflies (Diptera: Calliphoridae) of Fennoscandia and Denmark. *Fauna Entomologica Scandinavica* **24**: 7–272.

Ruzicka J. 1994. Seasonal activity and habitat associations of Silphidae and Leiodidae: Cholevinae (Coleoptera) in central Bohemia. *Acta Societatis Zoologicae Bohemoslovicae* **58**: 67–78.

Sadler D. W., Fuke C., Court F. and Pounder D. J. 1995. Drug accumulation in *Calliphora vicina* larvae. *Forensic Science International* **71**(3): 191–197.

Sadler D. W., Richardson J., Haigh S., Bruce G. and Pounder D. J. 1997. Amitriptyline accumulation and elimination in *Calliphora vicina* larvae. *American Journal of Forensic Medicine and Pathology* **18**(4): 397–403.

Saraste M. 1990. Structural features of cytochrome oxidase. *Quarterly Review of Biophysics* **23**: 331–366.

Schoenly K. and Reid W. 1987. Dynamics of heterotrophic succession in carrion arthropod assemblages: discrete series or continuum of change? *Oecologia* **73**: 192–202.

Schroeder H., Klotzbach H., Elias S., Augustin C. and Pueschel K. 2003. Use of PCR-RFLP for differentiation of calliphorid larvae (Diptera, Calliphoridae) on human corpses. *Forensic Science International* **132**: 76–81.

Shubeck P. P. 1968. Orientation of carrion beetles to carrion: random or non-random? *Journal of the New York Entomological Society* **76**: 253–265.

Simpson K. 1985. *Forensic Medicine*. Arnold: London.

Singh D. and Bharti M. 2001. Further observations on the nocturnal oviposition behaviour of blowflies (Diptera: Calliphoridae). *Forensic Science International* **120**: 124–126.

Smith K. G. V. 1975. The faunal succession of insects and other invertebrates on a dead fox. *Entomologist's Gazette* **26**: 277–287.

Smith K. G. V. 1986. *A Manual of Forensic Entomology*. The Trustees of the British Museum (Natural History): London.

Smith K. E. and Wall R. 1997. Asymmetric competition between larvae of the blowflies *Calliphora vicina* and *Lucilia sericata* in carrion. *Ecological Entomology* **22**: 468–474.

Smith K. E. and Wall R. 1998. Estimates of population density and dispersal in the blowfly *Lucilia sericata* (Diptera: Calliphoridae). *Bulletin of Entomological Research* **88**: 65–73.

Sperling F. A. H., Anderson G. S. and Hickey D. A. 1994. A DNA-based approach to the identification of insect species used for post-mortem interval estimation. *Journal of Forensic Sciences* **39**: 418–427.

Spitz W. (ed.). 1993. *Spitz and Fisher's Medico-legal Investigation of Death: Guidelines for the Application of Pathology to Crime Investigation*, 3rd edn. Charles C. Thomas: Springfield, IL.

Staerkeby M. 2001. Dead larvae of *Cynomya mortuorum* (L.) (Diptera: Calliphoridae) as indicators of the post-mortem interval – a case history from Norway. *Forensic Science International* **120**(1–2): 77–78.

Stevens J. 2003. The evolution of myiasis in blowflies (Calliphoridae). *International Journal of Parasitology* **33**(19): 1105–1113.

Stevens J. and Wall R. 1995. The use of random amplified polymorphic DNA (RAPD) analysis for studies of genetic variation in populations of the blowfly *Lucilia sericata* (Diptera: Calliphoridae) in southern England. *Bulletin of Entomological Research* **85**: 549–555.

Stevenson D. and Cocke J. 2000. *Integrated Pest Management of Flies in Texas Dairies*. Texas Agricultural Extension Service; The Texas A&M University System, B–6069: http://animalscience.tamu.edu/sub/academics/dairy/pubs.html; pp 1–30.

Stubbs A. and Chandler P. 1978. *A Dipterist's Handbook*. The Amateur Entomologist 15. The Amateur Entomologists' Society: London.

Tabor K. L., Brewster C. C. and Fell R. D. 2004. Analysis of the successional patterns of insects on carrion in southwest Virginia. *Journal of Medical Entomology* **41**(4): 785–795.

Tantawi T. I. and Greenberg B. 1993. The effects of killing and preserving solutions on estimates of larval age in forensic cases. *Journal of Forensic Sciences* **38**: 303–309.

Tantawi T. I., El-Kady E. M., Greenberg B. and El-Ghaffar H. A. 1996. Arthropod succession on exposed rabbit carrion in Alexandria, Egypt. *Journal of Medical Entomology* **33**(4): 566–580.

Toye S. A. 1970. Studies on the humidity and temperature reactions of *Dermestes maculatus* DeGeer (Coleoptera: Dermestidae) with reference to infestation in dried fish in Nigeria. *Bulletin Entomological Research* **60**: 23–31.

Turchetto M., Lafisca S. and Constantini G. 2001. Post mortem interval (PMI) determined by study of sarcophagous biocenoses: three cases from the province of Venice (Italy). *Forensic Science International* **120**: 28–31.

Unwin D. M. 1984. *A Key to the Families of British Diptera*. AIDGAP, Field Studies Council Publication No. S9 (now No. 143). Headly Brothers Ltd: London [reprinted from *Field Studies* **5**: 513–553 (1981)].

VanLaerhoven S. L. and Anderson G. S. 1996. Forensic entomology determining time of death in buried homicide victims using insect succession. Technical report TR–02–96. Canadian Police Research Centre: Ottawa, Ontario.

VanLaerhoven S. L. and Anderson G. S. 1999. Insect succession on buried carrion in two biogeoclimatic zones of British Columbia. *Journal of Forensic Sciences* **44**(1): 32–43.

Vass A. A. 2001. Beyond the grave – understanding human decomposition. *Microbiology Today* **28**(November): 190–192.

Vass A. A., Bass W. B., Wolt J. D., Foss J. E. and Ammons J. T. 1992. Time since death determinations of human cadavers using soil solution. *Journal of Forensic Sciences* **37**(5): 1236–1253.

Vass A. A., Smith R. R., Thompson C. V., Burnett M. N. *et al.* 2004. Decompositional odour analysis database. *Journal of Forensic Sciences* **49**(4): 1–10.

Vaz Nunes M. and Saunders D. S. 1989. The effect of larval temperature and photoperiod on the incidence of larval diapause in the blowfly, *Calliphora vicina*. *Physiological Entomology* **14**: 471–474.

Vincent S., Vian J. M. and Carlotti M. P. 2000. Partial sequence of the cytochrome oxidase *b* subunit gene I: a tool for the identification of European species of blowflies for post mortem interval estimation. *Journal of Forensic Sciences* **45**(4): 820–823.

Vinogradova E. B. and Marchenko M. I. 1984. The use of temperature parameters of fly growth in medico-legal practice. *Sudebno Meditsinskaya Ékspertiza* **27**: 16–19.

Wall R. 2004. Daily temperature fluctuation and the accumulation of day–degrees. Proceedings of the European Association for Forensic Entomology, 29–30 March 2004, London.

Wall R., French N. P. and Morgan K. L. 1993. Predicting the abundance of the sheep blowfly *Lucilia sericata* (Diptera: Calliphoridae). *Bulletin of Entomological Research* **83**: 431–436.

Wallman J. F. and Adams M. 2000. The forensic application of allozyme electrophoresis to the identification of blowfly larvae (Diptera: Calliphoridae) in southern Australia. *Journal of Forensic Sciences* **46**(3): 681–684.

Watson E. J. and Carlton C. E. 2005. Succession of forensically significant carrion beetle larvae on large carcasses (Coleoptera: Silphidae). *Southeastern Naturalist* **4**(2): 335–346.

Watson L. and Dallwitz M. J. 2003. *Onwards British Insects: the Families of Diptera*: http://delta-intkey.com [accessed 17 May 2005].

Wells J. D. and Sperling F. A. H. 1999. Molecular phylogeny of *Chrysomya albiceps* and *C. rufifacies* (Diptera: Calliphoridae). *Journal of Medical Entomology* **36**(3): 222–226.

Wells J. D. and Sperling F. A. H. 2001. DNA-based identification of forensically important Chrysomyinae (Diptera: Calliphoridae). *Forensic Science International* **120**: 110–115.

Wells J. D., Introna F., Di Vella G., Campobasso C. P. *et al.* 2001. Human and insect mitochondrial DNA analysis from maggots. *Journal of Forensic Sciences* **46**(3): 685–687.

Wigglesworth V. B. 1967. *The Principles of Insect Physiology*. Methuen: London.

Wilson L. T. and Barnett W. W. 1983. Degree–days: an aid in crop and pest management. *California Agriculture* **January–February**: 4–7.

Wilson D. S. and Fudge J. 1984. Burying beetles: intraspecific interactions and reproductive success in the field. *Ecological Entomology* **9**: 195–203.

Wilson D. S., Knollenberg W. G. and Fudge J. 1984. Species packing and temperature dependent competition among burying beetle (Silphidae, *Nicrophorus*). *Ecological Entomology* **9**: 205–216.

Wolff M., Uribe A., Ortiz A. and Duque P. 2001. A preliminary study of forensic entomology in Medellín, Colombia. *Forensic Science International* **120**(1–2): 53–59.

Wyss C. and Cherix D. 2002. Beyond the limits: the case of *Calliphora* species (Diptera, Calliphoridae). Proceedings of the First European Forensic Entomology Seminar, eafe.org/OISIN_2002; p 77.

Wyss C., Cherix D., Michaud K. and Romain N. 2003a. Pontes de *Calliphora vicina* Robineau-Desvoidy et de *Calliphora vomitoria* (Linné) (Diptères: Calliphoridae) sur un cadavre humain enseveli dans la neige. *Revue Internationale de Criminologie et de Police Technique et Scientifique* **1**: 112–116.

Wyss C., Chaubert S. and Cherix D. 2003b. Case study – determining post-mortem interval with four blowfly species (Diptera: Calliphoridae): the importance of cross-assessment. Proceedings of the European Association for Forensic Entomology Conference, 29–30 March 2003, London; p 28.

Zehner R., Häberle M. and Amendt J. 2004. Heteroplasmy in the *Necrobia* genome (Coleoptera: Cleridae): impact of DNA-based species determination. Proceedings of the European Association for Forensic Entomology Conference, 29–30 March 2004, London; p 28.

Zehner R., Armendt J., Schutt S., Sauer J. *et al.* 2004. Genetic identification of forensically important flesh flies (Diptera: Sarcophagidae). *International Journal of Legal Medicine* **118**(4): 245–247.

Zhong T., Chen Q., Wu R., Yao G. *et al.* 2002. The effect of extremely low frequency magnetic fields on cytochrome oxidase subunit I mRNA transcription. *Journal of Industrial Hygiene and Occupational Diseases* **20**(4): 249–251 [in Chinese].

Index